DR. STEV

NO OTHER GODS

GODS

THE

BIBLICAL

CREATION

WORLDVIEW

FOREWORD
DR. HENRY M. MORRIS

to grow in strength in the twenty first century, in my opinion, was the most important contribution to the conservative Christian movement in the last fifty years. There are now hundreds of creation organizations around the world.

Home schoolers and private Christian schools teach materials that have their roots in scientific creationism. There are thousands of pastors committed to preaching the word of God with authority from a literal creation worldview.

Thousands of scientists have been influenced by the evidences of science for creation that Dr. Morris so skillfully communicated and thoroughly documented over his long career.

Many of these have taken up the mantle and continue the ongoing discoveries that naturalists are no longer able to reject as religious wishful thinking.

What Dr. Henry M. Morris began, almost single handedly in the early days, by submitting his science to the guidance and submission of the word of God, has created a legacy that will continue to strengthen the conservative Christian movement for generations to come.

All of us who have been impacted by his books, the Institute for Creation Research, the debates, the seminars, and his prayers, are a part of that legacy.

I pray he fully enjoys his just reward for being faithful with what he was given, and may God continue to multiply his influence amongst us.

❧

"You ou shall have no other gods before Me." Exodus 20:3

❧

"Beware, lest your hearts be deceived and you turn away and serve other gods and worship them." Deuteronomy 11:16

❧

"What hat then? If some do not believe, their unbelief will not nullify the faithfulness of God, will it? May it never be! Rather, let God be found true, though every man a liar, as it is written, 'That thou mightest be justified in Thy words, And mightest prevail when Thou art judged.'" Romans 3:3-4

❧

INTRODUCTION

Statistics have shown that 70% of Christian young people entering college will no longer be committed to their faith by the time they graduate. What is the problem? Is it the fact that many of our Christian colleges and most of our state colleges employ professors who rejected their own faith in their college years and became committed to destroying the faith of the students who have come through their classes?

This certainly is a problem, but it is not the reason for the high drop out rate for young Christians losing their faith. No, the real problem is our young people are not being rooted and grounded in their faith by understanding their own worldview. In our churches, we often teach our children what to believe but we do not give them the solid reasons for why we believe what we believe.

They are not grounded in being able to give an answer for why they believe in the authority of scripture, why they accept it as a reliable account of history, why it is consistent with science, why they are convinced of its inerrancy, or why it is the one true worldview. Not having this foundation makes Christian young people, and adults for that matter, easy prey for professors committed to proselytizing students to their apostate opinions.

They pontificate about their modern and postmodern philosophical rationalizations (and irrationalizations) to their captive audiences while holding the power of giving grades over the heads of their answerless protege.

An uninformed church membership is the reason why so many young Christians lose interest and drop out of active involvement in their faith. Their church did not prepare them to give an answer to those who attack their faith.

When the Treasury Department trains their agents to identify counterfeit money, they do not show them any counterfeit bills. They teach them how to recognize every detail of genuine bills produced by the government. Once they know what genuine bills look like, they can distinguish them from counterfeit bills. My objective in writing this book is not to give summaries of all the counterfeit worldviews and philosophies there are in the world today.

My objective is to teach Christians the foundations of the Biblical creation worldview so they can recognize false teachings by knowing what genuine Biblical truth is. I not only want to communicate what that truth is, I want to give reasons why we know it is the truth. I will expose evolution theory as the foundation of all false worldviews of our day.

This book is a compilation of lectures given in the class, The Biblical Creation Worldview, I have been teaching for over twenty years. The three major texts used as required reading are the Bible, *The Genesis Record* (197jjfed6) by Dr. Henry M. Morris, and *Scientific Creationism* (1974) by Henry M. Morris.

The class covers the first eleven chapters of Genesis giving evidences from science, archeology, history, and scriptures gleaned from a wealth of sources supporting the Biblical creation worldview. It is offered as a tool to help ground Christians in the Biblical creation worldview so they can give an answer for what they believe.

The major thesis of the book is the Bible is its own worldview. It is written as a progressive revelation in history by eye witnesses. As a worldview, it shows itself to be consistent from beginning to end with its own understanding of reality. That understanding builds from the beginning and culminates in the historical person, Jesus Christ.

The bibliography in the back of the book includes every book I have read on this subject and gives several websites that can be used to read articles and find other books that are available. It is amazing how much information on creation has been published just in the 25 years I have been studying the Biblical creation worldview.

I believe there will come a day when Christian students will go into these classes that professors use to destroy young Christian's faith and actually be able to answer those professor's every argument. I believe those students will be used to turn some of those apostate professors back to faith in God. One day college campuses will become the place where many will finally come to the truth in the God of scripture. How will this happen?

Through churches that have pastors and teachers that are giving their members the truth about creation that stands against Satan's lies. I believe it because it is already happening. Evolution theory is coming apart at its foundations and creation theory is gaining more and more strength of evidence with each passing year. The church needs to catch up on this evidence and start getting the truth out. I pray this book will help.

Dr. Steve Kern

NO OTHER GODS - DR. STEVE KERN

CHAPTER 1
WHICH GOD?

WORLDVIEWS

Every person has a worldview. Dr. James W. Sire in his book, *The Universe Next Door*, in the fourth addition defines a worldview as, "A worldview is a commitment, a fundamental orientation of the heart, that can be expressed as a story or in a set of presuppositions (assumptions which may be true, partially true or entirely false) which we hold (consciously or subconsciously, consistently or inconsistently) about the basic constitution of reality, and that provides the foundation on which we live and move and have our being."[1]

There are as many types of worldviews as there are varieties of cultures in the world. There are even worldviews within major worldviews. A worldview can be defined more simply as each person's way of explaining their existence and everything around them. Your worldview tells you from where you have come, where you are headed, what is right, what is wrong, what is good, what is evil, what is important, what is not important, and so on. Religion or the lack of it plays a major role in defining a person's worldview.

Christianity, Islam, Hinduism, Buddhism, Judaism, Zorasterism, Bahi, Shinto, Tao, Voodoo, Shamanism, Communism, Secular Humanism, Atheism and all their many sub-sects are only a few of the major foundations of worldviews around the world.

The major objective of this book is finding and living the most accurate worldview. The assumption is that if there is a God, then He would have a worldview and has communicated it to us. We find that worldview documented in the Bible. A literal historical interpretation of the Bible is God's worldview revealed to us as His instructions on how to live in the reality of His absolute truth.

The Bible begins with a statement that in the twenty-first century raises a major question. The statement is, "In the beginning God created the heavens and the earth." The question is, "Which God?" Due to the break down of international barriers and the rise in cross-cultural interaction among the nations of the world, tolerance of all faiths, and the acceptance of pluralistic approaches to God have become the ideals of political, social, and religious correctness.

The growing worldwide acceptable answer to the question, "which God?," would be, all the gods, because they are all different manifestations of the same god. This god has revealed himself to different cultures in ways that would most readily lead people of that culture to worship him. The assumption is that there is no right or wrong worldview. Whatever worldview an individual becomes committed to is the right worldview for that person. It has nothing to do with truth. Instead, it has everything to do with personal preference. In the final analysis, this ideology allows a person to make up their own god, thus establishing themselves as their own god.

There are two major problems with this understanding that all religions lead to the same god. First, there is the principle called Truth. Truth tells us concepts that contradict each other cannot all be true.

In judicial court, there are often conflicting stories that have to be investigated to see which one is supported by the facts that can be presented. The contradictions are weeded out until the truth becomes established. In other words, truth does not contradict itself.

The Bible claims to be the truth of God. Jesus said in John 17:17, "...Thy word is truth." If the Bible reveals God as personal and transcendent, then He cannot be a pantheistic, impersonal, blended cosmic consciousness with the universe. Those are contradictory concepts and so, they both cannot be true. Pluralism works great for a "new world order," but it contradicts the principle of truth.

The second major problem with the plurality of religious paths to God is the God of the Bible makes a very intolerant declaration as the first of His Ten Commandments. This is what He commands in Exodus 20:3 , "You shall have no other gods before Me." This is a major problem for pluralists who believe all religions lead to the same god because the Biblical God reveals Himself in very specific terms about the kind of God He expects us to understand Him to be.

In Genesis chapters 1 - 11 we have God revealing Himself in very specific ways in a historical context. Genesis 1 is especially important because it defines Who God is and the kind of God He is specifically by the way He created. To get that first chapter wrong is to violate the first commandment. We need to understand that if Genesis is a historical record of God revealing Himself to man, then that record is absolute truth because events happening in history cannot be reversed once they take place (we will study absolute truth later).

If it is absolute truth, then any descriptions of God that contradict what God has revealed about Himself cannot be true because truth does not contradict itself. If God's revelation says the universe had a beginning in Genesis 1:1, then any religion that teaches the universe is eternal has to be talking about a different god. A universe that is eternal has to be its own creator while a universe with a beginning has to have a Creator. To worship any god that contradicts the God of the Bible's revelation of Himself is to worship false gods, which violates the first and foremost command of the Ten Commandments, "Thou shalt have no other gods before me."

This applies to science as well. The Bible teaches that God created in six literal twenty-four hour days and rested on the seventh day. If a, so called, scientific theory like evolution tries to teach that there was a "big bang" thirty billion years ago that began the universe and living forms developed from simple to complex forms from a common ancestor along the way through natural selection, then either the Bible is true or evolution theory is true but both cannot be true because they are opposite concepts.

The Bible teaches that God spoke the universe into existence, and formed each of the plants and animals according to their own kind in a six-literal-day process. This is not evolution. It does not matter if you try to invoke God as the supreme guide of the evolutionary process, it still contradicts the Biblical record. It also portrays God as a different god than the one described in Genesis 1, which violates the first commandment. An evolutionistic god is not the same God who gave us His revelation of His creative process that took place in six literal twenty-four hour days.

The two concepts leave you with a choice. Are you going to base your faith in the beginning of the universe on man's limited and continually changing assumptions determined by fallible scientific investigation, or are you going to place your faith in the revelation given by the infallible Creator of the universe? The question is this, which comes first, theology or science? We will study this further later. This study suggests that our theology should interpret our science rather than our science interpret our theology. The Biblical record makes it clear that evolution is bad science as well as a false understanding for interpreting Genesis. We need to let the Bible interpret itself as its own worldview and then use it to direct us in our scientific studies.

It has already been stated that worldviews have sub-sects within their own ranks. This is true of Christianity, Judaism, and Islam. All three faiths claim Genesis as a foundation of their faith. There are those in these three major monotheistic religions that believe Genesis teaches the earth is only a few thousand years old. There are others in these faiths that believe Genesis teaches the world is billions of years old.

The "short-agers" (who believe the earth is a few thousand years old) say the days of creation were six literal days with one day of rest to complete a literal creation week. On the other hand, "old-agers" (who believe the earth is four and a half billion years old) say the days symbolize long ages of evolutionistic or progressive creation events and we are now living in the seventh age of God resting from creating.

Here is the burden of this book. The God of the Bible has revealed what kind of Creator He is by the way He brought the universe into existence.

CHAPTER 1 - WHICH GOD?

The God who created in six literal days cannot be the same god who would create over billions of years. They are opposite concepts with many contradictions between each other and they both cannot be true. To misinterpret Genesis 1 is to violate the first commandment. It is actually a form of idolatry because misinterpretation leads to the worship of a false god out of a non-Biblical understanding of God. That is a very serious offense to the one true God. It is imperative that Bible scholars get the beginning of all revelation right. Not to do so is to establish a false foundation of understanding who the Biblical God is, thus leading people to worship a false god.

To get the book of Genesis right, especially the first eleven chapters, is to establish a theological position that will lead to getting the rest of God's worldview right. To do so is to heed Paul's exhortation for all teachers of God's word to, "Be diligent to present yourself approved of God as a workman who does not need to be ashamed, handling accurately the word of truth." (II Timothy 2:15). If not, then James 3:1 becomes ominous, "Let not many of you become teachers, my brethren, knowing that as such we shall incur a stricter judgment."

So, if there is a God, then the big question is, with all the worldviews competing with each other, which one has the true understanding of who God is? This study will attempt to show that the God of the Bible is the one true God who only can be worshiped correctly from a literal, historical interpretation of Genesis 1 -11. All other interpretations lead to a corrupted presentation of Who the true God really is. They create false impressions that distort God's revelation of Himself.

This distortion results in the worship of false gods that ultimately God will judge as Paul states in Romans 3:4, "May it never be! Rather, let God be found true, though every man be found a liar, as it is written, 'That Thou (God) mightest be justified in thy words, and mightest prevail when Thou art judged.'"

In other words, when the final judgment comes, God will show that His word has always been true regardless of man's continual distortions of it and all false teachings will be shown to be lies. God will then be justified to judge all those who chose to believe false worldviews and taught others to do the same. Thus, it is imperative that we get it right when it comes to having the right worldview and proper Biblical interpretation as a result of right understanding.

WHAT IS A THEORY?

In order to be confident that we have the right worldview, we need to define terms like theory, bias, and objective and subjective faith. We will define theory first. Scientist often try to make the point that the word "theory" is used differently in science terminology than in the mainstream. Science theory is understood to be supported by a great deal of reliable evidence. Science theory is often presented as the next best thing to a fact. In the final analysis, there is really very little difference between the two definitions.

Webster's New World Dictionary [2] defines a theory as, "a formulation of apparent relationships or underlying principles of certain observed phenomena which has been verified to some degree."

These principles began as hypotheses that were individually either proven or not. Notice the words "apparent" and "verified to some degree." There is no mention of the words "proven fact."

A theory can never be established as acceptable science if it is not observable, repeatable, and testable. Any components of a theory must be tested over and over again and give the same result at each testing in order to suggest its validity. It cannot be tested over and over if it is not repeatable.

There are different levels of science. There are hard science levels and soft science levels. Chemistry and physics are categorized as hard sciences because they are based more firmly on observations that are repeatable and testable. In contrast, the study of the origins of life in biology or the beginning of the universe in cosmology are considered to be areas of soft science because they depend more heavily on speculation and inference rather than actual observation. The beginnings of things are past events that are no longer taking place in the present. It is important to keep the differences between the two in mind as we continue our study.

When it comes to the study of worldviews and their different understandings of how the universe came into existence, they can only be understood as theories based on inference, with components that are not proven facts. Why is this so? It is not possible to travel back in time to observe how the universe began. All the events in that process can never be repeated and so they cannot be tested. Only the present is observable, repeatable, and testable. We can only speculate on beginnings by what information we have in the present that might shed some light on the past.

In this present time of understanding, there are those who try to declare their theory of beginnings as based on facts thought to be various proven hypotheses which are actually only evolutionistic assumptions.

To do this is scientifically dishonest because it is not possible to go back thirty billion years to observe if some big bang took place. They can only assume something took place possibly thirty billion years ago because they presently observe the stars and galaxies expanding away from the earth at a certain rate of speed, even an increasing rate of speed. These observations may suggest a cosmic explosion happened sometime in the past, but they do not prove it as fact.

The same is true of the creation worldview, whether understood in long ages or short ages of the earth. It can only be expressed as theory because it cannot be supported by scientific observation. It is not possible for us to travel back in time and watch how God created. We can only observe what is the results of that creative activity. There are several creation stories that have been passed down from antiquity and even those stories have different interpretations by various points of view, but none are based on present day scientific observation. Creationists then, like evolutionists, must present their understanding of beginnings as theory.

L. Harrison Matthews, in his *Introduction to Darwin's Theory,* makes the above point loud and clear that evolution theory and creation theory are both on equal footing. This is what he wrote, "The fact evolution is the backbone of biology, and biology is thus in the peculiar position of being a science founded on an unproven theory - is it then a science or a faith?

Belief in the theory of evolution is thus exactly parallel to belief in special creation - both are concepts which believers know to be true but neither, up to the present, has been capable of proof."[3]

The better way to compare the two opposite interpretations of beginnings is to present them as models. To compare the creation model with the evolution model, you must list their different assumptions. Those assumptions are then used to make predictions of what should be observed in the present if what the model's assumptions declare happened in the past are true.

The model that is confirmed by the most predictions without having to make major alterations to those predictions is more likely to be the closest to what actually happened in the past. Dr. Henry Morris wrote the book, *Scientific Creationism*, with the intent of showing how the creation model is far more supported by what is observed in the present than the evolution model. A study of that book will help you gain an understanding on how to think in terms of models of beginnings rather than theories.

There are several examples of how the creation model predicts what we observe where as the evolution model must adjust its theory to fit the observation. Take for instance the prediction of how animals and plants came into existence. According to the creation model, animals were created according to each of their own kind. Evolution says these organisms developed from simple one celled organisms into higher and higher levels of complexity. Creation predicts that DNA will be found to be genetically specific where as evolution would predict DNA will have the ability to transform from one genetic code to another.

What has been found is similarities between the DNA molecules of different life forms but no mechanism that is available to the DNA to make it add or delete information that would transform an organism from one kind into another. Actually, DNA contains error detecting and error correcting mechanisms to keep its information the same.

These mechanisms are designed to resist any change in the DNA. Mutations have been studied extensively as a possible transitional mechanism but they either take away information or add detrimental information, thus leading to a weakening of organisms rather than making it possible for them to transition into another more complex organism. Creation predicts this would be the case in order to maintain individual kinds. Evolution, on the other hand, must adjust to find another explanation of how animals evolved.

Another important prediction between creation and evolution would be organization. Evolution says that matter and living organisms evolved over billions of years by random processes governed by chance encounters of various forms of energy and matter in space and time. If this is true you would predict that the make up of the universe would be non-predictable with very loosely connected laws of operation with wide variables. Creation would predict that an intelligent, personal Creator would leave evidence of intelligence and design in the creation.

The fact of the matter is intelligence and design resulting from a personal Creator is found everywhere in the universe from the smallest cells to the largest galaxies.

Creation also predicts that structures in processes would all have to come into existence at the same moment in order to function effectively. This phenomenon is called "irreducible complexity." One such example is blood clotting. Without blood clotting, no living organism could survive its first injury. The clotting process is an intricate makeup of many factors being present at the same time which are totally dependent on each other to cause correct reactions that lead to clotting.

Creation predicts this would be the case. Evolution cannot explain how the process could work if all the parts involved in the process had to evolve separately over a long period of time. There are many physical structures and processes in living organisms that have been shown to be irreducibly complex as described by microbiologist Michel Behe in his book, *Darwin's Black Box*.

This irreducible complexity is important because it answers one of Charles Darwin's greatest concerns in his book, *On the Origin of Species*. Darwin admitted, "If it could be demonstrated that any complex organ existed, which could not possibly have been formed by numerous, successive, slight modifications, my theory would absolutely break down."[4] The fact is this irreducible complexity Darwin feared is found most especially in living organisms.

Logic itself makes it clear that every effect has to have a cause and the cause has to be greater than the effect. Actually, the evolution model that depends on chaos and random events provides no logical or philosophical base that would encourage scientific investigation in the first place. Science depends on predictability and establishing laws that can be proved by observation.

The evolution model based on naturalism does not predict predictability and laws of science because this worldview suggests nothing can really be known for sure if it came from a chaotic beginning with no guiding principles.

On the other hand, creationism tells us that the Creator is unchanging and established physical laws to govern the operations in the universe that can be studied and used to man's advantage, thus establishing a philosophical foundation for the establishment of scientific research. God actually commanded man to establish science in Genesis 1:28 when He told him to "subdue" the earth.

There are many predictions made by the young earth creation model that holds to a water canopy and a worldwide flood that are substantiated by what is observed in the present. Some examples are: Polar ice caps that now contain evidence that they were once warm tropical environments. A decaying electromagnetic field that could only be a few thousand years old according to its present decay rate, with a half life of 1400 years, it would be far too strong to allow life on earth just 10,000 years ago.

The elements such as salt, nickel, magnesium and many others accumulating in the oceans suggests they have been doing so for thousands of years not millions and billions. Worldwide rates of erosion suggest the mountains have been around only thousands of years. The fact that every place on earth gives evidence of having been under water in the past and that the geological formations are the result of huge catastrophic events are much better explained by a worldwide flood rather than many separate catastrophes over millions of years.

Evolution has to develop explanations to make these and many other observable realities fit their model. Creation actually predicts that these realities are exactly what you would expect to find if the creation model is true. This predictability makes it the stronger model by far adding strength to the validity of the creation theory.

UNDERSTANDING BIAS

Ken Ham, in the video series, *Answers In Genesis*, made it clear to me about how important it is that we understand the place of bias in our discussion of worldviews. Professional scientists often pride themselves in being adventurers in the pursuit of knowledge through their empirical investigations.

They believe they are open to accepting the results of their studies regardless of where they might take them. This being the case, they believe it is important not to allow any personal bias to distort their observations. But, intrinsically there is a problem involved with this honorable yet unrealistic ideal. It is not possible for any finite being to be unbiased.

Let me explain what I mean by first defining what bias is. The Webster's New World Dictionary defines bias as a, "mental leaning or inclination: partiality: prejudices." [5] What that means is every person is biased. As limited beings, we do not have complete knowledge. We are not capable of knowing the end from the beginning. There are many holes in our perception of how things are.

There is much we have to accept by faith because of our finite limitations to existence in time, space, and matter. Our finite predicament forces us to have leanings toward personal preferences of how we want things to be and those leanings incline us toward specific points of view. What I am describing here is the fact that every person has a worldview and your worldview will determine your bias.

Let me give you an example. If an astrophysicist believes evolution is true and the universe is old, then that person will automatically interpret what he or she observes from that specific assumption of his or her worldview even if a young universe explanation might be better. The old universe astrophysicist will not even consider the young universe explanation as plausible because it is too fantastic a thought for him or her to imagine. The same is true of astrophysicists who are creationists.

They may know and even understand the old universe explanations of cosmological phenomenon, and yet they will look for young universe explanations for these phenomenon because they are convinced they have the right worldview.

With these observations in mind, we must admit every person is biased and be honest about it. Bias, then, becomes just like our discussion about theories. We cannot remove bias from the debate over which is the best worldview, and so we must find a way to determine which bias is the best. That is why I am writing this book. I am willing to admit that I am biased by my young earth creation worldview and so I will give many of the reasons why I believe my bias is the best bias.

I believe I have solid theological, philosophical, historical, and scientific reasons for my bias.

Like theories describing beginnings are best compared by establishing models, so those models can be compared to determine which bias is supported by what can be observed. When you are involved in a debate with someone, you have to know what their worldview and their bias is. The only way to come to a consensus is for one side to be convinced to change their worldview and thus their bias. That is why a lot of debates never get anywhere because a person's bias is strongly entrenched in their belief system. It is something they hold to dearly because it defines their very existence. It is hard to admit you have been committed to a lie.

There is only one unbiased being in existence. That being has to be Creator God Himself. If God is infinite and totally unlimited in all the attributes of His being, then God has no reason to be biased. He knows all things from beginning to end. He does not have leanings toward one explanation or inclination to believe one theory or another.

He is the ultimate explanation of all things. He is the foundation of all truth. As I said in my explanations of worldviews, God is His own worldview and He has expressed how He wants us to understand that worldview and live within its truth. That worldview has been revealed to us as the Holy Scriptures. So, the issue is not whether you are biased or not. The real issue is, do you have the right bias? The scriptures are the key to finding the right bias based on what is really true.

BIBLICAL FAITH

Faith is a function of bias. We hold to our biases by limited knowledge and fill in the gaps with faith. Actually our biases are determined by our worldview and much of our worldview is accepted more by faith than by observable reality. Much of what we acknowledge as true is done much more by faith then by sight. Even so, that does not mean that our faith cannot be confirmed by being based on solid observation. There are two kinds of faith we find being expressed when it comes to examining worldviews. I refer to them as objective faith and subjective faith.

Objective faith is developed out of what a person can observe. There are those things we must believe even though we cannot observe them, but those things can be confirmed by what we can observe. For example, we believe in wind, electricity, and gravity not because we can see them but because we can observe what they do. The Bible tells us in Hebrews 11:6 that we must believe in God if we are going to be pleasing to Him. But, there is a problem here. The Bible also says in John 1:18 that, "No man has seen God at anytime..."

So, how can we have an objective faith in a God we cannot see? The answer is simple. We can observe those things that are in existence that confirm His existence. There are at least five major objective realities that can be observed that establish faith in the God of the Bible. They are the creation, the scriptures, the historical life of Christ, the Jews in history, and the Church which Jesus said that He would build. All five of these observable realities call for a supernatural explanation for their existence.

This book has been written to show how these five objective realities confirm the God of the Biblical worldview.

Then what about subjective faith? Subjective faith is a product of what a person determines to be true by what they discern from within themselves. All mystical Eastern religions are designed to help their adherents develop their own self-realization through forms of meditation to help them look within themselves for realty. If a religious or philosophical explanation sounds good while seeming to make sense of things and brings feelings of love, peace, and tranquility, then it must be true for that person who accepts it as true. These mystical worldviews do not trust what can be observed for they often view observable reality as an illusion.

For them, what can be experienced within is more real than what can be seen from without. Their understanding of God is an impersonal consciousness which can be experienced but not necessarily known. They have a pantheistic view of God suggesting the universe and all that it contains is god. Every part of the universe is god and so looking into oneself is to look for the part of god you are, thus achieving self-realization of ones own part of deity or the whole of universal cosmic consciousness.

According to the Biblical worldview, subjective faith grows as individuals are convicted by the person of the Holy Spirit concerning their sin and their need for God's Savior, Jesus Christ, the Son of God. When individuals yield to the Holy Spirit's conviction and repent of their sins while asking for forgiveness, we enter into a subjective relationship with God as the Holy Spirit enters into our inner most being.

Our subjective relationship with God producing love, joy, and peace begins to grow as the Holy Spirit renews our believing minds through the objective study of scripture and the other four major realities listed above that give understanding of God's purpose for us as His children.

God, in this relationship, is a personal being revealing Himself as a loving Father. Each of us enters into this personal relationship by choice emitting from our own personal free will. In contrast with mystical religions, Biblical subjective faith is the product of objective faith as we grow in our personal relationship with God. He is revealed in scripture as the God who is a person that can be known, at least, in a finite way as His Holy Spirit reveals Him to us. As we study scripture, the creation, the historical Christ, the history of the Jews, and the historical development of the Church made up of individual believers in Jesus Christ, our personal relationship with God grows.

Biblical faith is rooted in learning about objective truths that lead to logical conclusions. Instead of faith being a "leap into the dark," as Soren Kierkegaard's existential philosophy suggested, Biblical faith is a leap into the light. This light of God's truth shines like a lamp standing on a hill in the deep darkness of a cold moonless night.

The source of the light is the objective realities that lead us to the conviction that the Biblical worldview is the truth in the midst of all the lies.

The God of creation, Who created in order to have relationship with man, has made Himself knowable by faith through what He has revealed of Himself through His creation, His revealed word, the history of His chosen people, His personal coming to earth as Jesus Christ, and the continuing historical saga of His Church. Each of these observable realities, when studied in proper relationship, make it clear to those who want to see, that *Elohim* is the Creator and Jesus Christ is His Son.

In this book, we will see how the creation, the scriptures in Genesis 1-11, the history of the Jews, and Jesus Christ, when correlated with objective support of theology, philosophy, history, and science, provide a solid foundation to grow an ever maturing faith that produces a meaningful relationship with the God who created us for Himself.

The church throughout history, then, has always been made up of those who are living out that objective and subjective faith that the Holy Spirit has been confirming in the hearts of all those who have believed the testimony of Jesus Christ since the day of Pentecost.

CHAPTER 2
WHICH TESTIMONY?

THE HISTORICAL EYEWITNESS

Whether other worldviews want to recognize it or not, the literal, historical creationists have an argument that sets it apart from all other worldviews. It is an argument rooted in solid historical eyewitness testimony. The historical person, Jesus Christ, is pivotal to our understanding of how to interpret Genesis 1-11. Why is this so? No man in history ever made the claims Jesus Christ made and backed them up with prophetic fulfillment, eye-witnessed miracles, and teachings that transformed human history. What did He claim? He claimed to be the Son of God. In doing so, He made Himself equal with God. He made statements like, "I and the Father are one." "If you have seen Me, you have seen the Father." "Before Abraham was, I am." These statements as well as many others make it clear that Jesus Christ understood Himself to be a part of the Godhead.

The four gospels in the New Testament are the foundational historical records that give eyewitness accounts of what Jesus did, what He said, and Who He claimed to be. We have to ask ourselves, "Can the gospels be trusted to give an accurate account of what they claim about Jesus?" Josh McDowell in his little book, *More Than a Carpenter*, [6] outlines at least six criteria that are used by historians to determine if writings that they are investigating are historically sound. He shows how each of the four gospels satisfy all six of these criteria better than any other documents from the time of Christ or before.

The following is a list of those six criteria.

First, the gospels are geographically sound. All the names and places in the gospels have been shown to be exactly as they were in Jesus' day.

Second, most of the historical events recorded in the gospels are confirmed by other sources. Names like Herod, Augustus Caesar, and events like the census taken at Christ's birth are recorded in other historical documents of the time.

Third, they are also consistent with the culture of that time. One such example is the attitude of the Jews towards Samaritans.

Fourth, the language of the Gospels, Koine Greek with some translated Aramaic, is from that time period.

Fifth, there are several other writers of the time that attest to the gospels as being reliable. All the early church fathers refer to the four Gospels by quoting from them extensively, and attesting to their authorship as authentic.

Sixth, the Gospels and the other New Testament documents have parts of manuscripts that are only once removed from the autographic originals within 80 years.

The age of those manuscripts tell us that there was not nearly enough time from the life of Jesus to the writing of the gospels about His ministry to allow the accounts to transfer from eyewitness accounts to legends. We can also trust their accuracy by the fact that, up to this time, the text-wording of the original documents was unchanged in those manuscript copies of the originals passed down through the centuries.

What all this means is that the gospels are the most historically reliable records known to man. What they record about Jesus Christ can be trusted as valid information. Thus, Jesus' claim to be God come in the flesh should be taken very seriously as a claim backed up by very solid historically sound eyewitnesses. The point being that, if what Jesus claimed about Himself as equal to God is true, and we claim Him to be our Lord, then He must be our ultimate and final authority of what the truth is about all things, including creation. In other words, to claim to believe that Jesus is the Son of God and yet not to accept what He taught is a total contradiction.

Not only did Jesus claim to be God, His followers who wrote about Him were convinced that He was God who came to live amongst us as a man. The apostle John wrote of Jesus in his gospel, "In the beginning was the Word, and the Word was with God, and the Word was God," John 1:1. Peter spoke for all the disciples when he confessed who they believed Jesus was by his declaring, "You are the Christ, the Son of the Living God," in Matthew 16:16. Paul was probably the most specific about his conviction concerning Who he thought Jesus was when he wrote in Colossians 1:16-17, "For by Him all things were created, both in the heavens and on earth, visible and invisible, whether thrones or dominions or rulers or authorities, all things have been created by Him and for Him. And He is before all things, and in Him all things hold together."

Peter also begins his second letter with the following confession, "Simon Peter, a bond-servant and apostle of Jesus Christ, to those who have received a faith of the same kind as ours, by the righteousness of our God and Savior, Jesus Christ," II Peter 1:1. Peter actually refers to Jesus as God in this statement.

The above descriptions of Jesus as being with God and being God in the context of creation can be traced back to Proverbs 8:22-31 where wisdom is personified. Verse thirty actually says, "Then I was beside Him, as a master workman; and I was a daily delight, rejoicing always before Him..." This passage and others helped the New Testament writers understand that Jesus was the personification of the wisdom of God in His creative work in the body of a man. All the writers of the New Testament believed that Jesus was the Son of God and thus a part of the Godhead.

Why is this so important? The answer should be obvious. If Jesus Christ is the Creator come in the flesh, then He is the one historical eyewitness to the creation event. In Jesus Christ, you have a person of the triune Godhead who took part in the creation process. If this is true, and I believe it is, then what the historical Jesus taught about creation should be the final authority on how we believers in Jesus Christ are to understand the way God created. This will put us in a position to know that we are worshiping the right God revealed in Genesis. I am amazed at how so many teachers of the scriptures rely on the authority of man's limited scientific understanding rather than the teaching of Jesus Christ to determine their interpretation of Genesis. They have made an unreasonable choice when they allow man's science rather than God's revelation of the creation to be their final authority. That is amazing when you realize that man's science is constantly changing its position compared to the fact that God's revelation has been the same for the last six thousand years when it comes to describing the creation event.

I do not believe many "Christian" scholars understand their own obvious contradiction when they claim to believe Jesus Christ is the Son of God, and yet, teach Biblical interpretations that contradict what Jesus Himself taught.

That is exactly what they do though, when they teach most of the early chapters of Genesis as allegory and myth. For them, Adam and Eve were not historical people and the story of the fall was not a historical event. They say the flood was local, not worldwide, or never happened at all. They teach that the earth is billions of years old. They follow an evolutionistic understanding of the fossil record of animals coming into existence and then going extinct over millions of years, rather than God's creation of the plants and animals according to their own kind. All of these examples, as well as many others, directly contradict what Jesus said and did to express His understanding of Genesis.

This is equivalent to their saying, "I believe Jesus is the Son of God and I accept Him as my Lord, but I do not believe what He taught because it contradicts what I consider to be the true authority, man's science." The point is either Jesus, as the Son of God, is the final authority in all things or He is the authority in nothing. If we cannot trust Him to be right in one area of truth, how can we trust Him to be right in any other area of truth?

One major observation that can be made about Jesus is He was a literal, historical, creationist. We can see this in the following three ways. The way the gospels describe Jesus' background in relation to the book of Genesis, the way Jesus used the book of Genesis as authority for His teachings, and the miracles He performed. We will look at these three categories to show that Jesus had a Biblical, literal, historical, creationist worldview.

THE GOSPELS' DESCRIPTION OF JESUS

First, we find that the four Gospels' descriptions of Jesus give literal, historical references from the book of Genesis as their authority. We have already referred to John 1:1 where the apostle John establishes that Jesus was the *logos* (Word) of God in the beginning that refers back to Genesis 1:1 which says, "In the beginning God created the heavens and the Earth." The word "beginning" is an element of time and John, by stating in John 1:1, "In the beginning was the Word...," makes it clear that he understood that Jesus was a part of that original historical event.

Luke also makes the historical connection between Genesis and Jesus in the genealogy he records in Luke 3:23-38. We find there that Jesus and Joseph, Jesus' adopted father, are listed as historical persons right along with all those listed in Genesis 5, suggesting that Adam, Seth and the rest were just as much historical persons as Jesus, according to Luke. Luke's genealogy also suggests that there is a continual historical progression of time from Adam to Abraham, from Abraham to Moses, from Moses to David, and from David to Jesus that can be calculated as having only been about four thousand years. The virgin birth described in Matthew 1:25, after the genealogy of Jesus is listed, would be understood as a fulfillment of God's promise to defeat Satan (crushing the head of the serpent) through the seed of the woman in Genesis 3:15 four thousand years after the promise. Those four thousand years bring us to another important point that can be made here.

Many interpreters of the Bible try to use Psalm 90:4, a psalm of Moses, and II Peter 3:8 as proof texts to support the idea that the "days" in Genesis 1 can mean long periods of time rather than a literal twenty-four hour day.

These are the verses that tell us that a day to eternal God can be the same as a thousand years. There is a problem with this interpretation. It fails to recognize that those verses were written by men with a certain worldview (young earth creationists) that must be understood in order to know what they meant by what they wrote. Both Moses and Peter were literal historical creationists who understood earth history in thousands of years, not millions. They both had the same genealogical lists available to them in the time they lived. Moses had the lists in Genesis plus the lists of those living from the time of Joseph to his own time.

ARE WE IN THE 6,000 YRS OF WORLD HISTORY?

Moses understood that only about twenty-five hundred years had passed from the time Adam was created to his time. Peter had all the lists in the Old Testament plus those of his own family. He also knew of the kingly lists later recorded in Matthew 1 and Luke 3. Adding up all the numbers, Peter would have understood that about four thousand years had passed from the creation of Adam up to his time. (I am assuming Peter had his family list because there are two lists recorded for Joseph, Jesus' adopted father, one in Matthew 1 and the other in Luke 3. Joseph coming from the same economic level as Peter, working poor, suggests rich and poor families maintained their genealogies.)

II Peter 3:9 must also be considered when we want to understand what Peter meant when he said a day is as a thousand years to God in verse 8. Verse 9 says, "The Lord is not slow about His promises, as some count slowness." In this context, we see that Peter is making a point about how God is moving rapidly to fulfill His promise of redemption. What Peter is inferring is that it had only been four days, according to God's timetable (A day is a thousand years), from the time of the fall to the coming of Jesus as the promised Messiah.

With that inference in mind, we can say it has only been two days (2,000 years to man) since Jesus (as God) promised to return quickly in Revelation 22:12, according to God's time table. Actually, by Peter's use of the number thousand we must accept the fact that Peter looked at Earth's history up to his time in terms of about four thousand years, not millions let alone billions of years. The suggestion by old earth theologians that Peter, or even Moses for that matter, understood the six days in creation week were long periods of time is a perfect example of how Bible interpreters often interpret the Bible on the basis of their own personal bias rather than on the basis of what the Bible's worldview actually is. It is a case of men making the Bible writers say what the old earth theologians want them to say rather than letting the Bible writers say what they actually meant to say. It is obvious that, according to the genealogical time tables in the Bible, both Moses and Peter meant to say the creation was only a few thousand years old. Jesus, having memorized His own genealogies recorded in Matthew 1 and Luke 3, would have had the same understanding of the historical progress from the creation to His own time on earth as being 4000 years. Jesus saw them as inspired or He would not have allowed them to be a part of His word.

WHAT JESUS TAUGHT OF CREATION

Let us continue our three arguments that suggest Jesus was a literal, historical, creationist. We have already discussed the first argument which was that the gospels give information about Jesus that suggests what His worldview was.

The second argument involves the things Jesus actually said that tell us how Jesus interpreted the book of Genesis.

There are many examples of Jesus referring to Genesis, especially the first eleven chapters, in His teaching. We will see that these references to the Genesis account were given by Jesus as historical events that He used to support the authority of His teaching. John 5:45-47 establishes the authority of Moses' writings. We need to remember that Moses' writings in Jesus' time were understood to be the first five books of the Bible, called the Law, and included Genesis. These three verses quote Jesus as saying, "Do not think that I will accuse you before the Father; the one who accuses you is Moses, in whom you have set your hope. For if you believed Moses, you would believe Me; for He wrote of Me. But if you do not believe his writings, how will you believe my words?" Jesus' point here is the fact that if His listeners did not accept the authority of Moses, then they would never accept His teachings.

Jesus based all of His teaching on the authority of Moses' writings as being the inspired word of God. That being true, Jesus establishes that His own worldview was based on the worldview of Moses' writings. We must remember the first five books of the Bible established the history of the nation of Israel, through whom God promised the Messiah would be given to the world. The creation account, the fall, the flood, the tower of Babel, the covenants with Abraham, Isaac and Jacob, the Exodus, the giving of the Law on Mount Sinai, the wilderness wanderings are all a historical progression of God setting aside Israel to bring about His redemption of man through Jesus Christ. These are all a part of Moses' writings. By saying Moses wrote of Him, Jesus makes the point that He is the God who Moses wrote about.

He is the power of God's creative spoken word. He is the Son of God who would come through the seed of the woman. He is the promised seed of Abraham. He is the promised prophet God would send who would be greater than Moses found in Deuteronomy 18:15-19.

It is interesting to note that by claiming Moses as His authority, Jesus also established the authority of Moses as the one whom God has ordained as the accepted worldview. Jesus did this after His resurrection, which established Him as the Messiah. Jesus, as the Messiah, became the final authority on Biblical interpretation. As that authority, Jesus tells us that we must accept Moses' writings as the established authority of God's worldview and accept them as a historical progression rather than allegory and myth. Why must we accept them as history? Because that is how Jesus used them as His written authority. He taught them as historical accounts.

Jesus says something similar in John 5:45-47 and in Luke 24:44. This quote takes place after the resurrection and just before His ascension when He said, "These are My words which I spoke to you while I was still with you, that all things which were written about Me in the Law and the Prophets and the Psalms must be fulfilled." Once again, Jesus refers to the Law as being prophetic of His coming. Luke goes on to tell us that Jesus explained to His disciples how the Law, beginning with Genesis, and all the rest of the Old Testament were all about Him. His point was that His historical coming to earth was the fulfillment of God's progressive revelation through the history of Israel. The first coming of Christ has its greatest meaning when understood in the history of the scripture's past prophecies fulfilled and the expectations of His future second coming that are confirmed by what happened in the past. Because the Lord fulfilled the prophecies concerning His first coming, we can be confident He will fulfill His prophecies of His second coming in history.

Jesus refers to this historical progression of revelation in time in Mark 13:19. Jesus is speaking in the context of the coming Tribulation that will take place before His Second Coming. The verse says, "For those days will be a time of tribulation such as has not occurred since the beginning of the creation which God created, until now, and never shall." We have already seen that Jesus accepted Moses' writings as the authoritative explanation for how the creation began. It must be understood by His statement here in Mark 13:19 that Jesus accepted the time line that began in Genesis 1, the creation, and continued through the Old Testament up to His time in history and would continue until His return at the end of history.

John, in Revelation 22:13, quotes Jesus the risen, returning Lord as saying, "I am the Alpha and the Omega (the first and last letters of the Greek alphabet), the first and the last, the beginning and the end." What does Jesus mean by that statement? Surely He is implying that when you look at the beginning of time you will find Him there just as when you look at the end of time you find Jesus there as well. As God, Jesus Christ is the Lord of all history as it has been recorded in His word. That history began with Genesis 1:1, "In the beginning God created the heavens and the earth." By adding Revelation 22:20 where Jesus also says, "Yes, I am coming quickly." We can see His agreement with Peter who said in II Peter 3:9 that God is not slow about His promises by His promise here to come quickly. By quickly, Jesus would be inferring a few days, according to God's time table, and a few thousand years according to man's timetable, not millions or billions of years.

When Jesus was asked about divorce in Matthew 19:3-6 and Mark 10:6-7, He referred back to Genesis 1:27 and Genesis 2:24 as the historical authority for His answer. The important thing to recognize is that Jesus refers to the creation of Adam and Eve in the image of God and the statement in Genesis 2:24, "The two shall become one flesh" as historical events.

Their occurrence in time took place before the fall meaning that God established His standard for marriage as one man with one woman who were to be understood as a unity. It was man's fall into a sinful, selfish nature that jeopardized the standard, but God's original intent for marriage has never changed. Jesus obviously accepted these verses as historical events.

Absolute truth and authority are key here. Why is it so important that Genesis be seen as historical? The answer is that historical events establish absolute truth. Think of it like this. I got up this morning and had a bowl of oatmeal for breakfast. At this writing several hours have passed since that oatmeal eating event. There is now nothing that can happen to change the fact that I ate oatmeal for breakfast this morning. When I tell you I ate oatmeal for breakfast this morning, I am declaring absolute truth because the event is now established in history and is unchangeable. Historical events are absolute truth. As absolute truth, historical events have authority to establish what has happened in the past and directs in how to understand the present. The fact that God never changes helps us to know God with absolute confidence in the present by what He has done in past history that is absolute truth. The fact that Genesis records historical events establishes it as absolute truth.

That absolute truth is what gives it its authority (not to mention the fact that it is God's word). Jesus then, refers to the events of how God created Adam and Eve as His authority for His teaching about divorce. Those events established a historical precedent that, as far as Jesus was concerned in Matthew 19:3-6 and Mark 10:6-7, never changed in God's point of view. Jesus made several other comments that are drawn from Genesis 1-6.

He spoke of Satan's temptation of Eve (Genesis 3:1-5) in John 8:44 by saying, "You are of your father the devil, and you want to do the desires of your father. He was a murderer from the beginning, and does not stand in the truth, because there is no truth in him. Whenever he speaks a lie, he speaks from his own nature; for he is a liar, and the father of lies." Jesus is quoted in Matthew 23:35 and Luke 11:51 concerning the historical righteous blood of Abel. Jesus also referred to the flood as history in Noah's time in Matthew 24:38-39 and Luke 17:27. He used the flood event as a reminder of God's judgment that will once again be experienced by those living at the end of time. Jesus expresses each of these events in historical terms not as allegory or mythological stories.

WHAT JESUS DID TO SHOW HE WAS THE CREATOR

The miracles of Jesus are probably the strongest arguments that establish Jesus as the incarnate Creator of Genesis 1. The miracles are far better understood as a reflection of a literal, historical, twenty four hour day, six day, six thousand years ago creation than the old age evolutionistic or progressive creation interpretations. The God Jesus reveals Himself to be as the Son of God by His miracles, surely is consistent with the God revealed by a literal interpretation of the Genesis 1 creation account.

Before we get specifically into our discussion about the miracles, I want to answer the argument that some use to discredit the miracles as authentic. Some argue that God would never break His own natural laws (gravity, atomic make up of elements, space and time, chemicals relating to each other, etc.) by the use of miracles because that would undermine His intentions for establishing the physical laws in the first place.

To answer this argument, we need to explain the difference between, what I call, the three levels of law found in scripture. The three levels of law are natural law, ceremonial / civil law, and moral law. The natural laws were established by God to govern the operations of the physical universe. They are the things we study in science to discover how they work in relationship with each other. Some of these are things like water freezing at 32 degrees Fahrenheit, the difference between solids, liquids and gas, how fermentation works, gravity, electromagnetism and so on. These laws are temporary. They are a part of the temporary universe. They will one day be done away with when God brings in the new heaven and earth. God is totally sovereign over these laws and is the power source that actually makes them all work. God allows these laws to work in nature but He is not morally bound to do so. Miracles are God expressing His sovereignty over the temporary natural laws He created. The most obvious conclusion that can be made about Jesus doing miracles in the gospels is that Jesus wants us to understand He is the God of creation and thus, Lord over all the natural laws.

The second level of laws is the ceremonial / civil laws. These are the laws we find established most specifically in Exodus, Leviticus, Numbers, and summarized in Deuteronomy. God gives these laws to the Israelites to set them apart from the nations around them.

The ceremonial laws were given as prophetic practices that pointed to the coming Messiah. These were the laws of the tabernacle that included the sacrifices, the maintenance of the Holy Place and the Holy of Holies. They also included the holy days such as the Sabbaths and the annual feasts. All these ceremonial laws were fulfilled by Jesus as the Messiah.

For example, the sacrifices are no longer needed because Jesus' death on the cross became the ultimate sacrifice that all the ceremonial sacrifices pointed to. Jesus tells us in Matthew 5:17, "Do not think that I came to abolish the Law or the Prophets; I did not come to abolish, but to fulfill." The life and ministry of Jesus is the ultimate meaning of the ceremonial laws. The civil laws were given mainly to protect the Israelites from things that would weaken the social order of the nation. Laws about clean meats, leprosy, tithes and offerings, cities of refuge, tests for infidelity, clipping of the beard, tattoos etc. were designed to separate Israel from the practices of the pagans as well as keep the people healthy physically. Many of these civil laws are still practiced by Jews today as well as other people groups. Some are no longer necessary with the advent of Christianity. One example is the laws of eating clean and unclean animals.

While God was preparing Peter to take the gospel to Cornelius the Gentile, He told Peter to eat unclean four footed animals, crawling things and birds of the air. When Peter refused to eat what was unclean according to the civil law, God's response was, "What God has cleansed, no longer consider unholy." Apparently, God's point to Peter was that Jesus had removed the barriers between Jews and Gentiles through His death and resurrection. There was no longer a need for separation of the two.

When it comes to the moral laws, such as having no other gods, theft, murder, adultery, fornication, idolatry and respect for parents to name a few, these laws never change. These laws are the very expression of the character of Holy God. We find them established as a part of the original creation before the fall of Adam and Eve. The moral laws were given to uphold the sanctity of life, the sanctity of marriage, the sanctity of believing in the one true God, the sanctity of property and several others.

Not even God can do away with these laws because to do so would violate His own character. He would have to deny Himself for Who He is as being holy and just. To satisfy these moral laws, God had to pay the price His own law required for breaking them. That is why Jesus had to come to earth, live a sinless life, and die on the cross. The moral law requires that without the shedding of blood there is no removal of sin (Leviticus 5, and 6, and Hebrews 9:22). God had to buy the right to forgive our sins by providing the shed blood of His sinless Son, Jesus Christ. By satisfying His own moral law, God is now justified in giving forgiveness to all those who are willing to receive it.

Having this understanding of the three levels of law, we can now go on to show how Jesus expressed His Creatorhood through the miracles He performed. Those miracles help us to see that Jesus as God incarnate is sovereign over the natural laws, while fulfilling the ceremonial/civil laws and satisfying the moral law that required shed blood for redemption. The miracles of Jesus were done to show He was the Creator who had come in the flesh as the Son of God. The miracles were done to lead people to place their faith in Jesus as the promised Messiah so, in believing, they could be saved from the consequences of their sins.

Jesus makes this point in John 10:37-38 when He says, "If I do not do the works (miracles) of My Father, do not believe Me, but if I do them, though you do not believe Me, believe the works (miracles), that you may know and understand that the Father is in Me, and I in the Father."

All the miracles express the Lordship of Jesus over the creation. In John 6:21, the disciples experience the Lordship of Jesus over time, space, and matter.

In Genesis 1:1, God tells us He created the time, space, matter continuum by saying, "In the beginning (time), God created the heavens (space), and the earth (matter)." John 6:21 is a part of the story of Jesus coming to the disciples by walking on water. They had rowed about halfway across the lake in the midst of a storm when Jesus caught up with them. After getting over their immediate reaction of terror, they recognized Jesus and received Him into the boat. Then verse 21 records an astonishing statement, "... and immediately the boat was at the land to which they were going."

Now John was a fisherman who had fished on the sea of Galilee all of his life. He tells us they had rowed only three or four miles, meaning about halfway across this huge lake, knowing full well how long it took to row across the lake. It is clear that John wants us to understand that something out of the ordinary had taken place when they came to land as soon as Jesus got in the boat. It was a miracle in John's estimation. A miracle of time, space and, matter. The time is defined as "immediately." The space is defined by the fact that they were in the middle of the lake and the matter is defined by the boat and everything in it. In other words, God expressed His sovereignty over the dimensions of time, space, and matter by moving the people and everything in the boat from the middle of the lake to the shore instantly.

Psalm 107: 23-30 is, in a way, prophetic of this miracle where it describes men in a raging sea that God calms and brings the ship safely to harbor. Surely a man like John, who made his living on the Sea of Galilee as a fisherman, would remember this Psalm in light of what Jesus did that stormy night and be reminded that only God can calm storms and move boats immediately to safety. John tells us he and the other disciples experienced Psalm 107: 23-30 literally.

The Lordship of Jesus over the dimensions of time, space and matter is also expressed by Jesus' ability, after the resurrection, to enter a room while the door remained shut. Obviously, Jesus had the power to move in other dimensions that go beyond the limits of time, space, and matter. The study of particles in quantum mechanics suggests there are other dimensions beyond time, space, and matter. This is very consistent with the fact that God exists in a realm called "spiritual" that supercedes the dimensions of our created universe. Genesis 1:1 says, "In the beginning God," meaning God had to have existed in other dimensions before time, space, and matter were created. Jesus in His resurrected body, being God, would naturally have the ability to move in this realm that certainly is made up of many more dimensions than those we are limited to in our finite existence. It makes us wonder what kind of dimensions we will move in as a part of the new heaven and earth with resurrected bodies like Jesus now has.

Revelation 21 and 22 tell us that God will dwell with us and there will no longer be a need for light producers like the sun because God Himself will be our light. Think of it. God, as the person Jesus Christ, came to live with us in our domain so He could take those, who choose to go, to live with Him in His spirit domain.

A domain no longer limited by time, space, and matter. Genesis 1:3 quotes God as saying, "Let there be light.." Verse 4 goes on to say, "...and God separated the light from the darkness..." When Jesus was hanging on the cross, we are told that it became dark for 3 hours (Luke 23:44). That would not have been a local event because verse 45 says the sun was "obscured." Josh McDowell in his classic book, *Evidence that Demands a Verdict*, on page 84 gives two historical references during the time of Jesus by Thallus and Phlegon that refer to an unexplained eclipse during the time of a full moon.

This occurrence would be consistent with the sun standing still in Joshua's time and the moving of the sundial's shadow back in the time of King Hezekiah. It is also consistent with the fourth day of creation when God created the sun, moon, and stars. If you can believe Genesis 1:1, "In the beginning God created the heavens and the earth," you can believe this account of 3 hours of darkness. It is reasonable to assume that in the darkest hours of man's history the Creator would cause a deep darkness to fall on earth to signify the magnitude of that darkest event.

We see the Lordship of Jesus over the elements of nature in several miracles. The miracle of changing water into wine suggests Christ's sovereignty over the whole process. He created the water, the nutrients in the earth, the vines that produce the grapes from those nutrients, as well as the sunlight that provides the energy for the vine to make the chemical processes happen in the vine. He also created the process of fermentation. The miracle was that Jesus put all those processes together at the same time to change the water into wine. By those at the wedding saying it was the best wine, they affirmed that God does all things well.

The miracles of the feeding of the five thousand and the four thousand also reveal the Lord's sovereignty over the elements and processes of nature. We learn by these miracles that Jesus had the power to recreate the bread over and over again, as He divided each of the small loaves. It is a definite reflection on His not only being the Creator of the process of making bread from wheat, but also showing that He is the same God who fed Elijah, the widow of Zarephath, and her son out of the bowl of flour and the jar of oil that did not run out until God sent rain. (I Kings 17:8-16) It also reflects on II Kings 4:42-44 where Elisha fed 100 people with little food and had plenty left over.

Of course these miracles also remind us that God created fish on the fifth day of creation and has the power to recreate fish, even in their cooked state. In John 21:9, I believe the fish cooking on the fire with bread already prepared by the resurrected Jesus, when He appeared to the disciples, after they had been fishing all night, was a reminder to the disciples of the loaves and the fish miracles to help them know it was truly Him. He was also reminding them of His ability to meet their needs when they became the fishers of men He called them to become. They had no need to go back to fishing again. Jesus was going to take care of them from then on.

The Lord also revealed His power over the forces of nature in several ways. He spoke to the wind and it became calm, which is a reminder of Genesis 8:1 where it says, "... and God caused a wind to pass over the earth, and the water subsided." Wind is caused by the coming together of cold and hot air suggesting God has the power to heat and cool air at will according to His own will. Gravity is also a force of nature and yet it had no effect on Jesus when He walked on water and ascended into heaven in bodily form.

The natural laws of how liquid and solids react with each other had no power over Jesus unless He allowed them to, as in His baptism when He was immersed in the water. His sovereignty not only applied to Himself in this miracle of walking on water, because we are told that He made it possible for Peter to walk on water as well. There is one miracle that seems to be done in a whimsical way by Jesus at first sight, until you allow it to be understood in the context of scripture. I am referring to the withering of the fig tree. Trees take several years to grow and a long time to wither but when Jesus spoke to this tree, it withered instantly according to Matthew 21:19. What seems to be whimsical is the fact that it was not the season for figs, according to Mark 11:13.

Why would Jesus curse a tree that was not suppose to be producing fruit? I believe there are two answers rooted in the context of the Biblical worldview. First, with the water canopy in place before the flood (which we will study later), the growing season for fruit trees would have been year-round. After God's judgment by the flood, the canopy was lost and thus, reduced the growing of fruit to a season rather than year round. The fig tree, without fruit, would be a reminder to the Creator of the fall and the judgment of the flood that finally altered the creation from its original beauty and productivity. Second, we find that the withering tree is a reflection on God's planting trees in the garden of Eden that would have grown from seed to maturity instantaneously.

The God who has the authority to make trees grow instantly also has the power to make them wither instantly. We see Jesus, through the withered fig tree, revealing Himself to be the Son of God Who is the Creator Who creates, grows, and withers fig trees. He is the same God who planted the garden in Genesis 2.

In the gospels, we are told that Jesus raised at least three people from the dead besides being resurrected Himself. There was the widow's son, who was being taken to be buried, when Jesus stopped the funeral procession and gave the young man back to his mother. Jairus was a synagogue official whose young daughter had died but Jesus gave her back to her father alive. Lazarus was a close friend of Jesus, but Jesus waited four days before He raised his friend from the dead and gave him back to his two sisters. These miracles over death take us back to the fifth and six days of creation when God created the animals with life and then man with life in the image of God. They also point to the curse of death in Genesis 3 where God pronounced death on man and all creation because of Adam's sin.

These miracles of resurrection power make it clear that Jesus is Lord over life and death. One of the great things we learn about Jesus is the compassion and empathy He shows for those who are sick, diseased, and crippled. This would be the kind of response you would expect from the Creator who originally made all things very good and then had to curse it all because of man's being deceived into rebelling against Him. If evolution were the Creator's means of bringing higher and higher forms of life into existence through pain and death, then to show compassion by violating His own creative process would be seen as contradictory to His integrity by intervening in the creative process He had established.

His compassion and empathy would suggest the evolution process of survival of the fittest through death, dying, pain, and suffering was not really acceptable by His own admission. By healing the sick, diseased, and crippled, Jesus makes it clear that these results of the curse were not a part of His original creative processes nor were they a part of His original intent to make everything very good.

Jesus also establishes the fact that He intends for His new heaven and earth to be free of disease and physical defects. That brings up a point to ponder. If God is a god of evolutionary processes, what makes us think heaven will be any better than what life is like now? "Old earthers" cannot answer that question honestly without admitting that Jesus does not fit the evolutionistic idea of how God created using death, dying, pain, suffering, and survival of the fittest mechanisms. The compassion and mercy of Jesus healing miracles is a direct contradiction to evolutionistic ideologies. There are three different accounts that speak of Jesus having Lordship over the animals.

There is an instance at the beginning of His ministry that Jesus used a catch of fish to show His authority over animals. His future disciples had fished all night and not caught a thing. Jesus tells them to cast on the other side of the boat. They obeyed and they could hardly pull their nets in because there were so many fish. In response, Peter fell to his knees and said in Luke 5:8, "Depart from me, for I am a sinful man, O Lord!" Peter's response makes it clear that he understood that he was in the presence of a man of God who did miracles. Jesus did the same miracle again, after His resurrection, for Peter and the other disciples to show He was the same Lord that had done this same miracle when they had first been convinced to follow Jesus. The third miracle has to do with a fish as well.

In Matthew 17:24-27 a tax collector had come to Peter to collect a tax. Neither Jesus nor the disciples had any money to pay the tax so Jesus sent Peter to go catch a fish with a coin in its mouth.

This miracle tells us many things about the total control God has over His creation. First, He knew where the coin was on the bottom of the Sea of Galilee. It was probably dropped there by a fisherman or lost in a storm when a boat capsized. Second, He knew which fish to send to pick up the coin. Third, He knew where the fish needed to be in order for Peter to be able to catch it and retrieve the coin. The miracle tells us God knows where everything is. We learn from it that God has total control over the animal kingdom and has power to use them for His purposes. It tells us God has creative ways to meet the needs of His people when they will trust Him. It also confirms the story of Jonah that presents a large fish sent by God to swallow a man and deliver him right where he was supposed to be, just as the fish brought the coin to Peter.

And finally, the fish and coin makes the point that God expects his people to be subject to the government by paying their taxes. Why is this important? Because, Genesis 9:6 suggests and Paul confirms in Romans 13 that God establishes governments to protect man from man. That being the case, every citizen under a government's authority should pay their part to make sure their government has what it needs to do its job.

All angels are created beings, including Satan (Ezekiel 28:13) and his fellow fallen angels, better known as demons. That being true, then you would expect even demons to be in subjection to their Creator, in spite of their own rebellion against that Creator. This is exactly what is communicated in the gospel records, such as Mark 1:27, "...He commands even the unclean spirits (demons), and they obey Him."

When Jesus approached the Gadarean demoniac, the demons possessing the man referred to Jesus as the Son of God, in Matthew 8:29.

This is important in that demons are spiritual beings and they knew who Jesus was from a spiritual perspective. By confessing Him to be the Son of God and obeying His commands, the demons make it clear that Jesus was their higher authority as a part of the Godhead and confirm that He was preexistent with God as John 1:1-5, Colossians 1:15-17, and Hebrews 1:1-2 state. When the angels were created, the Son of God was the power source of God's creation of them. One of the most obvious miracles that points to Jesus as the Creator is the healing of a blind man found in John 9. Jesus first refers to Himself as "the light of the world" in verse 5, which is a reminder of Genesis 1:3 where God says, "let there be light" and I John 1:5 that declares, "...God is light, and in Him there is no darkness at all."

Jesus then, does something unusual. It is unusual because Jesus would often speak and a person was healed or Jesus would just touch them and they would be healed. In this account of healing a blind man from birth, He spits on the ground and makes some mud from the dirt and wipes it on the blind man's eyes. He then tells the man to go and wash in the pool of Siloam. The blind man obeys and receives his sight. Here is the point. Jesus making mud from the dust of the ground was an obvious reference to the fact that God formed man's body, including his eyes, out of the dust of the ground according to Genesis 2:7. By doing this, He was showing that He is the God who created the eyes of man from the dust of the ground. The washing of the eyes in the pool can also be a reference to the fact that the ground came out of water on the third day of creation.

One final indicator of Jesus being the Creator described in Genesis 1 is found in the Gospel of Mark more than the other gospels, although each refers to this phenomenon.

I am speaking of the word "immediately" that is used over and over again in Mark. Everything Jesus did was immediately. If God created everything in the universe in six literal days, then most of that creative activity would have to have taken place immediately. The Biblical understanding of how God works is quickly not through slow processes of long periods of time. We have already referred to this in our discussion of Peter's explanation of how a day is as a thousand years to God. He is not slow to bring about His promises. That is what you would expect of a God who created the universe in six literal days, not billions of years. He would have done His creating immediately. Jesus was a literal, historical, creationist. His genealogy in Luke 3 connects Him to Adam as His historical descendant.

As the Messiah, He certainly is regarded as the fulfillment of Genesis 3:15; a verse that foreshadows the virgin birth that Matthew 1:25 describes. Much of the authority of His teaching was rooted and grounded in a historical interpretation of Genesis. All of His miracles are much more reasonably understood to have been expressions of the power of the God we find in a literal understanding of Genesis 1 and 2 rather than an allegorical, evolutionistic, or progressive creation interpretation. With these conclusions in mind, Jesus is established as the final authority in interpreting Genesis 1 and 2. If Jesus was a literal creationist, then following Him as our final authority gives a strong argument to those who accept His worldview as the one that is true. The bottom line for believers in Jesus Christ as the Son of God and the Creator expressed in human flesh ultimately is that we should be literal creationists because Jesus was a literal creationist. If that is what Jesus taught, then that is what we should believe because Jesus, the Creator incarnate, is our final authority.

CHAPTER 3
WHICH BEGINNING?

BIBLICAL CONSISTENCY

We have already seen that the Bible is its own unique worldview. I have made the point that the scriptures give us God's revelation of how He expects us to understand reality by making God's worldview our worldview. If the Bible is its own worldview, then it will be consistent with its own presentation of reality and rely on itself as its own authority for interpretation. That was one of the parameters for accepting books as a part of the canon of scripture. Those books had to be consistent with the books that had come before them as inspired writings by men of God whose writings were obvious additions to God's progressive revelation of Himself and His redemptive plan through history. The standard for that consistency of revelation was established by the book of Genesis. Genesis is the beginning of all revelation from God. This being true all the rest of the books in scripture depend on that beginning for their own interpretation.

Here is the problem. All through history, and especially in this day in time, many, so called "Bible scholars," have interpreted scripture from various worldviews trying to superimpose assumptions from their worldviews onto the scripture's worldview. In so doing, they suggest contradictions in the scriptures, because their worldview's assumptions contradict the scripture's worldview. For example, some scholars assume the earth is billions of years old, as secular scientists have been claiming for the last century and a half.

They assume the geological formations of the earth formed slowly over long periods of time. That being the case, they must reinterpret the flood in Genesis 7, as a local flood, or define the story as a myth without historical basis. But, the language of Genesis 7 strongly describes a global flood that was experienced by eye witnesses. These "old earth" Bible interpreters must then manipulate the language by expanding the meaning of words to make them fit their worldview rather than allowing the language of Genesis to say what the original writer intended to say. Their worldview contradicts the Biblical worldview and so they must rearrange the meaning of words in the scriptures to fit their unbiblical assumptions. This then causes contradictions in the minds of many who read the Bible in light of these "old earth" scholar's forced interpretations.

The beautiful thing about scripture is that it does not contradict itself when you allow it to be its own worldview. It is totally consistent with its own theology, history, science, and understanding of man and his fallen dilemma (anthropology). It is only when other worldviews are forced on it that the Bible loses its continuity. Herein lies the challenge. We should be allowing God's word to tell us who we are rather than our trying to tell it what we want it to say. The study of scripture should be a continual evaluation of ourselves as to whether every dimension of our worldview is a product of God's worldview or just one that has its roots in our culture and secular philosophies we learned in school, gained from continual mass advertising, or absorbed from movies and television. It is amazing how everything starts to make sense, when we just let the word of God say what it says and rearrange our way of thinking to the way God has told us reality actually is. You discover that God's word is dependable, authoritative, and can be trusted to lead you into all truth.

That is my own testimony. I grew up in church and went to public schools. I studied the Bible in Sunday School and studied science at public school. I experienced a subtle but very real frustration as a young person trying to reconcile in my mind how the Bible taught that God created everything in six literal days about six thousand years ago, while man's science was telling me the universe started by a big bang billions of years ago — two opposite concepts. The evolution based interpretations of Genesis always seemed like a stretch to me when I read what the Biblical account actually said. It was a mighty liberation for me, when I finally learned from Bible believing scientists that man's science had got it all wrong and that I could just believe God's word for what it says.

Since that time, my confidence in the scriptures, as the revealed word of God, has been affirmed over and over again. It all fits. It all makes perfect sense to me now and I can honestly concur with Paul in I Corinthians 14:33 that, "God is not the author of confusion." I am totally convinced that God is totally capable of saying what He means and meaning what He says. God did not have to wait on us for 6,000 years to figure out evolution before we could finally understand His word. God's word from the beginning has always said exactly what God intended for it to say, and it has always been right on.

The Bible being consistent with itself as a historical record of God progressively revealing Himself over time is all about objective faith. As we allow the scriptures to say what they say and then begin to learn all the wonderful information in science (such as the fossil record) and archeology that support the literal creation model given in scripture, our faith is strengthened and our confidence in God's word begins to grow.

Our apologetics, all of a sudden, become based in observable reality and we can support our faith with what we know is true, not what we hope is true. It is now time for us to get into the study of what God's word actually says and why we should make it our worldview.

THE HISTORICAL RECORD

Any worldview that accepts the idea that earth and the universe is billions of years old has to reject theology or science that teaches they are young. Theistic evolutionists, age day theorists, progressive creationists, or any new age cult using the Bible as a part of the mix in their teaching, has to relegate the creation account to allegorical or mythical literature. They suggest the Genesis 1:1 - 2:3 account is poetry much like other creation stories from other ancient cultures. This is necessary because the events described in Genesis 1:1- 2:3 happening in a matter of seven days, twenty-four hour days, six thousand years ago is totally unacceptable to their billions of years mindset. But an important question to ask is, in the original language is this account written, as poetry is written in the rest of the Bible, or is it written as historical narrative as in the rest of the Bible?

The Institute in Creation Research (ICR, founded by Dr. Henry Morris) did an eight year study called, Radioisotopes and the Age of the Earth, or RATE.

One of the team members was a Hebrew scholar, Dr. Steven Boyd, who was asked to do a literary study of Genesis 1:1 -2:3 to answer the above question. The results of his study are recorded in chapter 10 of the RATE report book, *Thousands... Not Billions*, compiled by Dr. Don DeYoung[7].

The RATE team was looking for evidence in the creation that would show old-earth radioactive decay dating methods to be wrong. As a part of their study, they wanted to be sure that the Genesis record itself could be shown to be a historical narrative. This would support one of their basic assumptions that the word of God can be trusted to be true as a historical record of how God created. If it is a true historical record, then radio isotope dating that gives billions of years' dates would have to be flawed.

Dr. Boyd's major approach to the study was to define the literary differences between Hebrew poetry and Hebrew historical narrative. He did this by comparing the use of verbs in several Biblical passages accepted by scholars as poetry and other passages accepted as historical narratives. He used a total of 97 texts, 48 being narrative and 49 being poetry. The Genesis 1:1 - 2:3 account was included in the study to see where it might fall in the midst of the results. The four major verb forms used were preterite, imperfect, perfect, and *Waw*-perfect. The preterite verbs were found to be used most profoundly in the narrative texts, whereas the imperfects and perfects were used mostly in poetry. The Genesis 1:1 -2:3 passage was found to fall very high on the narrative side of the different graphs used to report the study's findings. The results of Dr. Boyd's study made it very clear that the creation account was written as a historical narrative, not as poetry.

The only honest way to understand the Genesis account of creation is to see it as a historical narrative. To turn it into a poetical allegory or mythological poetry does violence to the scriptures. It gives a false understanding of the intent of the writer (probably God Himself) to give an eyewitness account of the week long creation event.

The Genesis 1:1 - 2:3 account is not literally satisfying unless it is read as a historical narrative. Trying to make it into something it is not calls for interpretive gyrations that were never intended by the author. The best understanding of the text is to just let it say what it says as a straight forward historical narrative. Genesis 1:1 - 2:3 is definitely not Hebrew poetry.

A BOOK OF BOOKS

Genesis is an interesting book. It could be said to be a prophetic expression of how the rest of the Bible would be organized, as it came into existence. The real power of this fact expresses to us the consistency of God's word with itself. Just as the Bible is a collection of 66 books written by eyewitnesses inspired by God over a period of four thousand years and finally compiled into one book by godly men, Genesis is structured in the same way as a compilation of eleven books written over a period of two thousand years by eyewitnesses inspired by God and finally compiled into one book by Moses, the man of God.

Because of its supreme importance as the foundation of the rest of God's progressive revelation of Himself through history, the authenticity of Genesis has been under attack more than any other book in the Bible. Most of the attack has come from liberal scholars (using higher criticism) with an evolutionistic worldview who suggest that much of Genesis, especially the first eleven chapters, came after the polytheistic Babylonian and Egyptian creation and flood stories. They theorize that the monotheistic view of one God in Genesis is a later product of an evolving development of man's perception of the older polytheistic gods.

The Documentary Hypothesis was their major explanation for the origins of the Genesis stories. This theory of Biblical interpretation attempted to turn Genesis into a patchwork quilt of mythological stories that were derived from long lost writings they defined as the Jehovahist, Elohimist, Priestly, and Deuteronomist documents. Their theorizing was total speculation of the Hebrew and Greek Old testament texts that over time has proven to be unfounded in light of over a hundred years of archeological findings in the Near East.

One of the liberal scholar's major speculations was that Moses could not have written Genesis let alone the Pentateuch (the first five books of the Bible), because writing had not developed until the time of the dispersion of Israel into Babylon. That assumption was totally destroyed when libraries that predated Abraham were unearthed in Palestine and the Mesopotamian Valley. One such library was unearthed in the Ur of the Chaldees, the very city that Abraham left to go to the land of Canaan. Of course, the Egyptians have long since been known to have writing far earlier than the time of the Raamses pharaohs. The Bible tells us the Exodus, led by Moses, took place at the time of the building of the storage cities Pithom and Raamses in early Egyptian history.

This attempt to remove Moses as the author of the Genesis record, as well as the rest of the Pentateuch, was a subtle way of trying to remove the historical authenticity of these important Biblical records. Their goal was to knock the Bible off of its high pedestal of unique historical authority, in order to have it viewed as nothing more than another human attempt among many to explain the origins of all things.

There has never been a lot of doubt in the ranks of conservative scholars that Moses was the author of the first five books of the Bible. The Bible itself, extra-Biblical writings such as the Talmud, and of course Jesus, all agree that Moses wrote the Pentateuch. But we learn from reading the book of Genesis, that it has an altogether different make-up compared to Moses' other four books in the Bible. Whereas Moses' last four books move naturally from beginning to end as one continuing dissertation, Genesis gives credit to other sources other than Moses that suggest Genesis is a compilation of several records joined together by an editor or compiler of those previous historical accounts. The Old Testament is made up of 39 books written over thousands of years that were later canonized into a book of books. Those books were canonized by godly men who believed those books fit their criteria for God inspired works. The same was true of the books in the New Testament canonized by leaders in the churches. It seems Moses, the man of God, had available to him preexisting writings that He saw as inspired by God and compiled them into the first canon of scripture, Genesis.

In reference to preexistent writing, Dr. D. James Kennedy in his book, *Why I Believe*, gives a report on archeological finds in Babylon and Acadia of thousands of written tablets written long before Abraham.

On one such tablet, a Babylonian king writes about how much he enjoyed reading the writings of those who lived before the flood. Another tablet lists the names of the ten kings of Babylon before the flood. It then mentions "the deluge" (flood), then continues the list after the flood. This makes it clear there were records before the flood carried over after the flood. There must have been a pre-flood Babylon, just as there was a pre-flood Assyria (Genesis 2:14).

The interesting point is Noah was ten generations after Adam, just as there were ten Babylonian kings before the flood. This tablet agrees with the time frame of ten generations the Bible also records before the flood.[8]

What evidence is there in Genesis that supports the claim that Genesis is a compilation of writings by authors who lived before Moses and had their writings passed down to Moses? Commentary writers of the book of Genesis call them *toledah* after the same Hebrew word that is translated "generations" or "origins" which can be expanded to mean "records of the origins." There are eleven of these *toledoth* statements in the book of Genesis. Each one states who the eyewitness was that recorded the information in the writing that precedes the *toledoth*. One such toledoth is found in Genesis 5:1, "This is the book of the generations of Adam." This statement best fits the information found in the previous record that goes back to Genesis 2:4, the first *toledoth*. Much of the information from Genesis 2:4 to Genesis 5:1 is information that only Adam would have been an eyewitness to. All the other *toledah* include the name of the one who would have recorded the previous events as an eyewitness to those events.

This strengthens the assumption that Genesis is a record given by men who experienced the events and establishes Genesis as a historical document with eyewitness authority, just as the rest of the books in the Bible are written in a historical context. One final note. The first *toledoth* in Genesis 2:4 says, "These are the generations of the heavens and of the earth when they were created, in the day that the Lord God made the earth and the heavens." (KJV). Interestingly, it does not give the name of a human eyewitness like the other ten *toledah*.

Of course, God was the only one present during the creation week. Adam and Eve were the last of God's creative work being created on the sixth day after all the rest of His creative work was done. The first *toledoth* tells us that God Himself would have given this written record to Adam as the first recorded scripture for Adam and Eve to learn about God from. It is awesome to realize that God has given each generation from the beginning a revelation of Himself so that, when it comes to knowing God, all of mankind is without excuse.

IN THE BEGINNING

Dr. Henry Morris made a statement in his book, *The Genesis Record*, that after reading it, I have never forgotten. The statement was, "...if a person really believes Genesis 1:1, he will not find it difficult to believe anything else recorded in the Bible."[9] What does it say? "In the beginning, God created the heavens and the earth." That is the crux of the whole matter of accepting the Biblical worldview. If you can believe a transcendent, infinite, all-knowing, all-powerful God created the universe and everything in it at the beginning of time, then the rest of God's progressive revelation of Himself and His redemptive plan falls into place.

You will have no problem with a six day creation. Actually, you will wonder why He took a whole week when He could have done it all at once. You will have no problem with believing in a worldwide flood, miracles, fulfilled prophecy, angels, the devil, demons, God becoming a man, or anything else you find in scripture because they are all consistent with the sovereignty of God as Creator, Sustainer, Ruler, and Redeemer of His creation.

The statement, "In the beginning, God..." tells us a lot about the God who is not introduced here, but stated as the historical fact of reality even before time began. It tells us that God is transcendent to His creation. This means that before the universe was created, God had always existed. That tells us that the universe is finite in time and God is infinite in time. The universe had a beginning and that God, who had no beginning, is its beginner. It is the classic statement of the law of cause and effect. This rule of science tells us that for every effect there has to be a cause and the cause has to be greater than the effect. Genesis 1:1 establishes this law by making it clear that the universe and all it contains is an effect and God, its Creator, is the great Cause of it all.

The word, "transcendent," refers to an important concept. We have already seen how it means that God existed before the universe. But, it means much more. It tells us that the universe is totally dependent on God. The universe exists because God exists. We can rightfully say, in light of God's transcendency, that God can exist without the universe but the universe cannot exist without God. The transcendence of God nullifies the concept of pantheism held by many eastern religions and New Age philosophies which teach the universe is god and god is the universe.

An example would be Hindu gurus who seek to help their followers find the god within themselves. Self-actualization is the process of becoming one with the universe or finding the god consciousness that is already in you as a part of the universal being. No, this is wrong, the universe teaches us a lot about its Creator but the creation is not its own creator.

We find God as He reveals Himself to us as a Person seeking a personal relationship with us through His creation, His word, His Son, and those people who have already come to know Him as a person.

God, being a person brings up another point of God's transcendency. If God is the First Cause of all things, then all of our qualities of being a person are effects that come from God. Our very own attributes of being individuals who posses personhood as an effect of the First Cause tells us our Creator has to be a person who has transferred His attributes of personhood to us. That would tell us that our need for relationship, acceptance, love, helping each other and the inward drive to know God are all effects that come from our Creator who is a person Himself. We will see later that, through Adam and Eve, God made all of us in His image or likeness. That means God made us with attributes similar to His that would allow us to finitely appreciate who God is and all that God is from a limited but meaningful perspective. For example, God has put the capacity within us to love our children. That love for our children comes from our Creator who loves His created children far more than we can comprehend. My point is that if we are products of chance random events, then an emotion such as love for offspring has no cause or even reason for its existence. A loving Creator, Who calls Himself a Father, is this parental emotion's only plausible explanation.

God existing before the universe gives us another very important insight to our existence. God's transcendence tells us what the foundation of all truth is. If God is the beginner of all creation, then all truth begins with God. God Himself, actually, is the Truth. In God then, all truth is absolute for God is absolute. This truth is expressed in the saying, "He is the same yesterday, today, and tomorrow."

In other words, God never changes and neither does the truth. Remove God, as evolution theory tries to do, and truth becomes relative. Truth loses its authority. But, God is absolute and so His truth is absolute. Dr. Gene Veith, in his book, *Post Modern Times*, makes this very point, "God " not culture " is the origin of meaning, truth, and values. As the Author of existence, God is authoritative. Thus, certain absolute truths and transcendent values are universal in their scope and application."[10]

Jesus said in His high priestly prayer in John 17:17, "Thy word is truth." Why is God's word truth? It is truth because God Himself is the truth. All truth begins with Genesis 1:1, "In the beginning God created the heavens and the earth." That statement, being true, becomes the foundation of all theology, all philosophy, all history, all science, our understanding about everything. It also means that Genesis 1, which defines the statement of how God created, gives the true parameters of understanding all these disciplines.

We need to keep in mind that Jesus said in John 4:24, "God is Spirit, and those who worship Him must worship Him in spirit and in truth." What that tells us is the spiritual dimension or dimensions existed before the physical dimensions of the creation and so what is really true is first found in the spiritual realm.

To leave the boundaries of the physical universe is to enter the spiritual dimensions because God's transcendency tells us that the physical universe is being stretched out by God in the midst of dimensions that supersede time, space, and matter and are called, by some, hyperspace.

The study of particles in atoms and quantum mechanics has continued to boggle naturalistic scientists' minds for years now because their activity definitely points to the existence of other dimensions beyond the physical. Truth, then, extends far beyond what we can experience in our physical capacities. That there is far more to our existence than meets the physical eye, is what it all means. What we experience in our physical existence is only a small part of all that is really true.

THE FOUNDATION OF UNDERSTANDING

Genesis 1:1, " In the beginning God created the heavens and the earth," is the foundation for understanding every discipline of study man has endeavored to create. It is the foundation of the study of theology, history, philosophy, and the many categories of studies that fall under the major category of science. If Genesis 1:1 is true, than all understanding begins with accepting the fact that God created the universe and all it contains. Theology becomes the study of the one true God revealed in scripture beginning with Genesis 1:1. History becomes the study of events that have been happening in the progress of time beginning with Genesis 1:1. The study of philosophy becomes simplified because all the great questions of our existence like, Who am I? Where did I come from? and What is the meaning of life? are answered for us in Genesis.

Science is learning how creation operates by learning the God established laws which govern nature. As we learn how God's creation works, we can bring it under our control for the good of all mankind.

It can also expand our understanding of God's creative work so our worship and praise of Him can become more and more filled with awe for our mighty Creator God.

Because of our culture's continuing infatuation with the scientific method and the great strides that are being made in new discoveries, science and secular scientists have been elevated to the position of supreme judges. In the minds of secularists, they alone must be depended on to lead us into all knowledge and understanding. But, Genesis 1:1 forces us to ask a question, "which comes first, theology or science?" To state it another way, every one of us has to make a rudimentary decision that will determine how we succeed in our quest for finding the truth. The decision is, am I going to use theology to interpret my science or am I going to use science to interpret my theology. To make my decision, I had to answer another question. Which came first, theology or science? Genesis 1:1 provides the answer for me. God existed long before science was ever necessary. As a matter of fact, God, as the Creator of the universe, created all the processes that man's science is still trying to figure out. That tells me I need to be right about what I understand about God if I ever hope to get science right. In other words, you have to have good theology to have good science.

Here is a huge problem. Science, as the final authority, is being used today as the backdrop for the study and interpretation of the Bible rather than God's word being used as the backdrop for guiding our scientific studies. Here is what I mean. Man's fallible and often times bad science, tries to tell us the earth is billions of years old. Scholars, who accept this to be true then go to Genesis and change the meaning of words like "day" in Genesis 1 to mean long periods of time.

For these "old earth" scholars, their science influences the way they understand their theology. On the other hand, a scholar that allows the Bible to direct his study of science will look for indicators in the creation that suggest the earth is only a few thousand years old. The two major examples of these two approaches are Dr. Henry Morris and Dr. Hugh Ross. Both men are evangelical Christians. Dr. Morris, as a "young earth" creation scientist (hydrologist), starts with the Bible and allows it to guide him in his scientific studies. On the other hand, Hugh Ross, an "old earth" progressive creationist and astrophysicist, allows his understanding of science to guide him, as he interprets Genesis 1. To Dr. Ross, the days of Genesis 1 are long periods of time joined together by creative acts that God performed over millions of years. He teaches this because that is what his secular science education has taught him, and he has accepted it as his final authority.

Here is our choice. Is the Bible our final authority or is man's science our final authority? I believe the Bible is God's word and God does not make mistakes like man does. God's Word has been around since the beginning of creation. I believe I will stick with God's word as my guide to lead me into finding the truth. It is as Paul said in II Timothy 3:16-17, "All scripture is inspired by God and profitable for teaching, for reproof, for correction, for training in righteousness that the man of God may be adequate, equipped for every good work." That verse tells me that to be the best theologian or the best scientist you can be, God's Word must be your starting point. God's word begins with Genesis 1.

INTELLIGENT DESIGN

An effect requires a cause. A beginning requires a beginner. A design also requires a designer.

These are all logical conclusions based on observation in the natural order of things. There are several specific arguments for design that are scientifically based. Evolutionists have tried to resist the Intelligent Design movement infiltrating debates in science circles by claiming there is no way to support Intelligent Design arguments from purely scientific explanations. To evolutionists, Intelligent Design is creationism in disguise. Resist as they may, there are at least three strong Intelligent Design scientific arguments dealing with DNA alone that are observable and falsifiable. First, DNA has encoded information in logical sequences that require meticulous organization. Information does not happen by chance. Second, the information is not explained by physical or chemical laws. Information has to have a source. Third, the probability of these codes coming into existence by chance is a resounding zero.

Other arguments include the formation of the first living cell. Even the simplest of cells are so complex the mathematical probability of one forming by chance is impossible. Besides that, any atmosphere containing oxygen would destroy amino acids before they could ever form randomly. As we have seen earlier, there are many mechanisms in living cells and physical bodies that are irreducibly complex and could not have evolved from simple to complex operations. The Cambrian explosion (all levels of life-forms found together in the lowest age of the geological column) confirms all life-forms began to exist at the same time.

These and many other observable evidences have convinced thousands of scientists, including many that are not creationists, to conclude that natural evolution cannot explain these phenomena.

These scientists, though often unwilling to identify the Intelligence, are convinced there had to have been a guiding intelligence involved in the process. Genesis 1:1 has been teaching Intelligent Design for thousands of years.

Intelligent Design can be a friend or enemy of the Biblical creation worldview. Many of those pushing the Intelligent Design movement are not interested in defining the Designer. For scientists, that is a philosophical question (Who is the designer?) that goes beyond the scope of science. Every religion, except atheism, believes in a designer or force that directs natural processes. Hindus, Buddhists, Animists, Polytheists, Muslims, Jews, Christians all believe in a Designer. It is up to those who believe the Biblical Creator is the Designer to present their case in the most convincing way. Intelligent Design only opens the door to discuss design, it does not commit to any one religious view. It is up to those of us who accept the creation worldview to fill in the gaps about Who the true Creator is. Intelligent Design does not have that objective.

GENESIS 1 STRUCTURE

We have already seen the structure of the book of Genesis broken down into eleven *toledah*. Genesis 1 also has some interesting aspects to its structure. We will look at four of these aspects of structure that are made obvious by the use of the words "and", "good," "day," and "created."

The first word we will consider is the Hebrew word *waw* which is translated into the conjunction "and" in English. In the Hebrew text of Genesis 1, every verse begins with *waw* except verse 1.

By the use of this grammatic structure, the Hebrew writer makes it clear that the reader should understand that a continuing, unbroken process is being described. The actual feel of the reading of the whole chapter is one long sentence. This is a strong evidence against the Gap Theory that suggests a gap of long periods of time took place between verse 1 and verse 2. The context suggests a continuing flow, and to insert a gap does violence to the text. The *waw* conjunction also supports the idea of a continuing process of creative activity that has a beginning and an ending without long slow intervals of time. Remember, the original text was not written with chapter and verse divisions. The original was one continuing text. With this in mind, the "ands" that begin verse 2 and 3 in chapter 2 also suggest that the seventh day was to be understood as the final day in the creation week. God's rest was to be included as the finish of God's work. This makes it clear that the seventh day falls under the same category of day as the six days described in chapter 1. This brings into question the interpretation that day seven is an allegorical period of time that we are still living in as suggested by some progressive creationists or theistic evolutionists.

The next word that suggests a structure is the word "good" translated from the Hebrew word *tobe*. *Tobe* is the word in Hebrew that has the broadest applications for referring to what is good. Jesus makes an important inference to the importance of the use of the word "good" when He responded to the rich, young ruler. The young man, who impressed Jesus with his sincerity, approached Him with a form of flattery by addressing Jesus as, "Good Teacher..." in Mark 10:17.

Jesus' response is interesting in verse 18 when He asked, "Why do you call me good? No one is good except God alone."

There are two points to be made in light of Jesus' question. First, by telling us only God is good, it seems that Jesus is asking the young man if he was saying that he believed Jesus was God. That would be consistent with Jesus leading the young man to salvation faith that requires that we believe Jesus is the Christ, the Son of the living God, just as Peter confessed in Matthew 16:16. Secondly, the fact that only God is good would tell us that the use of the word "good" in Genesis 1 tells us that each of those levels of good were perfect expressions of the good Creator God. The creation was good because God is good.

There are seven declarations of "good" in Genesis 1. It is important to note that not all of those declarations come at the end of a day of creative activity. Instead, they seem to be declared at points where a predetermined phase has been completed suggesting a plan that is being followed. It is as though God declares each phase to be exactly what He intended it to be and thus "good" or a perfect expression of Himself as the Creator, Who is good. You might compare it to the idea of how an artist's work says a lot about the artist, and God certainly intended for us to learn a lot about Him by the study of His creation. It was a perfect work created by the infinite Artist.

There is one expression of "good" that seems to be an exception to this phase explanation of structure to Genesis 1. That "good" declaration is the first one found in verse 4 where it says, "And God saw that the light was good..." This particular "good" seems to refer more to the phenomena of light in creation rather than a phase.

This is an important distinction to make when we learn in the rest of the Bible that light and God are inner connected.

The apostle John tells us in 1 John 1:5, "...God is light, and in Him there is no darkness at all." This statement takes on greater meaning when we learn in Revelation 21:23 about the new Jerusalem and how it will be lighted, "And the city has no need of the sun or of the moon to shine upon it, for the glory of God has illumined it, and its lamp is the Lamb (Jesus Christ)." Revelation 22:5 reiterates this system of lighting by saying, "And there shall no longer be any night; and they shall not have need of the light of a lamp nor light of the sun, because the Lord shall illumine them; and they shall reign forever and ever." These verses tell us that the light of God is not just symbolic of God's truth but that the light of God has the capacity to give needed light in whatever He creates. As we look at Genesis 1 we see there that the light spoken of in verse 4 is not a light being generated by the sun because the sun is not created until the fourth day in verse 14. This suggests, to me, that the light spoken of in Genesis 1:4 and Revelation 21:23 and 22:5 are the same light. This is confirmed even more when God Himself states that the light is "good" and Jesus has already told us that only God is good.

I believe what this first declaration of "good" is telling us that God Himself enters into the confines of His creation to provide His light that will be the source of all life. John 1:4 actually infers this in reverse order where it says of Jesus, "In Him was life, and the life was the light of men." Here the light of God and life are used interchangeably. Think of it. God, who exists in dimensions of spirit that surrounds the sphere of the limits of our universe's dimensions of time, space, and matter, entered the created universe to provide His needed light so we could share His life.

More than that, God entered our domain so He could finally bring those who love Him into His domain to live with Him for eternity. And so, this first declaration of "good" defines for us the light as being God's light for only God is good.

There is another subtle but very real truth established in verse 4 where God separates the light from the darkness. We will learn later that God's ultimate intention for us all is to choose to live in the light or live in the darkness. At the end of time we see that God finally separates those who chose in this physical life to love the darkness and those who chose to love the light. Those who chose the darkness will spend eternity in the lake of fire. That final Hell will be a place of darkness, worms, and gnashing of teeth as Jesus tells us in Matthew 22:13. Those who chose to love the light and live in that light will spend eternity in the new heaven and earth where there is no darkness at all. We are told in Revelation 22:5 that there is no night there. From the beginning, God has been separating the light from the darkness by choosing those who choose to love Him and by rejecting those who choose to reject His love. Instead, they love the evil that emits from the darkness of their own depraved hearts rather than the goodness of God.

The other six declarations of "good" then define for us the phases of creation that God had predetermined as His plan to follow in His six days of creative activity. Each phase was necessary to complete in order to enter the next phase, but each phase was a perfect expression of the kind of good God our Creator is. The first phase established the time, space, matter continuum that would provide the finite place and building blocks of unenergized particles in verse 1.

The necessary sources of energy of the Spirit's vibrations and the light of God are provided in verse 2 to begin the process of forming particles into elements that form the first level of matter, which was water that took on the form of a sphere. Within that sphere, other elements were energized into gases that would make up the needed atmosphere that would provide the breath of life. This atmosphere comes on day 2. Out of the sphere of water under the atmosphere, God then energized the elements to become the foundational land of the earth and all the minerals necessary to provide nourishment required to sustain the physical bodies He would create out of those same elements. So on the first half of the third day, God completed the land and seas that were to become the support system for the rest of His creation on earth. This finished the first phase that included the providing of the "good light" and allowed for the second declaration of "good."

The second phase and third declaration of "good" took place in only a half of a day. This phase included the making of all vegetation out of the elements that will transform the already made minerals into food. This food would be in place for the living creatures that would be created on the fifth and sixth days. This vegetation would include that growing in the newly formed seas and that growing on the newly formed land formed in the first phase. The third phase and fourth declaration of "good" took a whole day and included the forming of the sun, moon, planets, and stars. These were to be light bearers and instruments for the use of keeping time. They were formed by the particles floating in space that were energized into balls of fire called stars, or huge rocks called planets, and moons floating in space around the planets.

The stars would produce light and the planets and moons would reflect that light.

The fourth phase and fifth declaration of "good" included the creation of living sea animals and birds that took place on the fifth day and the fifth phase. The sixth declaration of "good" included the creation of all land animals on the first half of the sixth day. The sixth and final phase was the creation of man (Adam and Eve) and culminated with a final declaration that it was all "very good." This "very good" signified a final completion of God's creative plan that included all the phases but also suggested that the creation of man was the crowning point of all His creative work. The creation of man declared as "very good" suggested that man was what the whole creative process was all about. In other words, God's original intent was to create Adam and Eve but He went through a whole process of creating an environment that was suitable for them. The creation was also designed to teach them about God and provide a place where they could have a relationship with God. But more than all of this, it was "very good" because it was all a perfect expression of the Creator God, Who is good.

The third word that is used to develop another structure in Genesis 1 is "*yom*" the Hebrew word translated into English as "day." Although the phases of creation are not limited to each day in the creation week, it is obvious that the writer wants us to understand that God Created in six days and rested on the seventh day, thus establishing the basic measurement of time as a cycle of seven day weeks. It is interesting to learn that all calendars in every culture are based on a seven day week. There have been countries that tried to change their calendars to six or eight days of a week rather than seven days but finally reverted back to seven days.

A seven day week seems to fit our physical make-up best for work and rest. The seven day creation week in Genesis 1 is the only place in antiquity where we find the establishment of a seven day week as the breakdown of a month or year. The lunar orbit around the earth establishes the measurements for a month. The orbit of the earth around the sun is how we measure a year. We also measure a day by the earth's rotation in relationship to the sun. One rotation equals one day. There is nothing that defines for us a seven day week except the Genesis 1 account. This suggests that the Genesis account is the foundation of all other culture's understanding of a seven day week. This strongly suggests it to be the original creation account.

God creating in six days and resting on the seventh day is the most basic structure established in Genesis 1. The first impression one gets when reading Genesis 1 is that the writer is wanting the readers of this account to understand it is talking about six literal days. The only reason to make it say something else would be the interpreter has a worldview that calls for long periods of time. Because of this "old earth" bias, he must force the word "day" to adhere to that worldview that violates the context of a literal reading. There are several reasons why the word day should be understood as a literal twenty four hour day. First, the Hebrew "*yom*" in scripture always means a literal twenty four hour day or the time of light in a day. This is true in 98 % of its occurrences unless the context makes it clear that the word is suggesting a period of time.

Even when suggesting periods of time *yom* only refers to short periods of time like a few days, a few years, or a hundred years or so but never millions of years.

But by far, *yom* is understood to be a literal day. The second reason why "day" should be understood as a literal day is because the days in Genesis 1 are all defined as the time of light in verse 5 when God, "called the light day and the darkness He called night." Third, "day" is further defined as having a literal evening and morning. The fourth reason is the word "day" in scripture, when preceded by a numeral one, two, three, and so on, is always understood to mean a literal day, not a long period of time. Fifth, all Hebrew lexicons committed to the traditional interpretation of the Old Testament define *yom* in Genesis 1 as a literal 24 hour day. Only lexicons influenced by Classical Greek philosophy question the literal day in Genesis 1. This is true of the early church fathers as well. Except for the Alexandrian School, all the early church fathers were literalists. It seems that the writer of Genesis 1 went to great lengths to make sure "day" was understood to mean a literal day and those first seven days made up the first literal week of history.

The fourth word that suggests a specific structure in Genesis 1 is the word "created" that is translated from the Hebrew word *bara*. This is a very important word in scripture because it is used only in reference to God. Only God has the power to call into existence those things that did not exist. He also has the power to make something no longer exist that He called into existence as we read in Revelation 21:1, "...for the first heaven and earth passed away (vanished)." *Bara'* was only used three times in Genesis 1. Each use suggested a new level of creative activity.

The first level is found in Genesis 1:1 where we are told, "In the beginning God created the heavens and the earth."

This first level of creation provided the space and building blocks (matter) related to time. He will use these dimensions to make the physical universe out of. By building blocks, I am referring to the particles that are energized to make up the atoms that then combine to make up the elements of matter. God created the building blocks out of nothing and then energized them to make the water, the atmosphere, the land and seas, the vegetation, the lights in space, and the physical bodies of animals and man.

The second use of *bara'* is found in verse 21 that says, "and God created the great sea monsters..." This new level of creative activity brought into existence all living beings. They were sea creatures, flying creatures, and land dwelling creatures that all possessed "life," or the Hebrew *nephish* in their bodies. This "life" separated these creatures from all non-living things that were previously created to support life. Their physical bodies were formed out of the same elements as the water, atmosphere, land, and vegetation, but in such a way that they had the capacity to posses life and have some level of awareness of their existence. "The life is in the blood," we are told in Leviticus 17:11, and so the Bible makes a distinction between living things who posses blood where as plants posses sap but not blood. It is important to note that none of the vegetation has "life." Vegetation was made out of the elements to support "life," but vegetation does not posses life as defined by Genesis 1.

The third *bara'* is found in verse 27 where it says, "And God created man in His own image." This ultimate *bara'* tells us that man was created on a whole new level separated above the other two levels.

We see this higher level of creation defined in Genesis 2:7 where we are told that man was formed individually out of the dust of the ground (animals were created in hosts, Genesis 2:1) and God breathed into man the very life and Spirit of God. This tells us that human beings are not descendants of animals but descendants from God. Man was not created a little lower than the angels but a little lower than God Himself, as we are told in Psalm 8:5, and is rightly translated in the New American Standard Bible. It is interesting to note that, like the animals, the angels were created as a host of angels. We will see this later. Likewise the angels did not have God breath into their bodies the very Spirit of God like He did for Adam. And so, each *bara'* describes a new level of creative activity with the creation of man being the highest level of God's creative work.

So, Genesis 1 is highly structured. That structure suggests to us that God was following a predetermined plan of creative activity that truly developed from simple to more and more complex arrangements of the building blocks (particles to atoms to elements to matter) He started with in verse 1 as opposed to evolving from simple to complex life forms. That should tell us that God has a plan and that He is still working out that plan. God never does anything hap-hazard but He works instead with purpose and predetermined outcomes. We are definitely not a cosmic mistake, products of random meaningless events. Ultimately, man was God's main objective as He went through the creation process. Providing a suitable place for God and man to have fellowship is what the creation is all about.

THE CREATION OF ANGELS

This would be an appropriate place to discuss when the angels were created. Those holding to the Gap Theory believe that angels, such as Lucifer, ruled a previous earth that was destroyed by God when Lucifer and a third of the angels rebelled against God, thus becoming Satan and the demonic forces of this present creation. They understand Isaiah 14:12-17 and Ezekiel 28:12-19 as having taken place before the recreation of the earth. But there is a great deal of Biblical evidence that suggest that the angels, including Lucifer, were created during the first three days of creation. They were to be ministering spirits for God and faithful believers. Hebrews 1:14 says of angels, "Are they all not ministering spirits, sent out to render service for the sake of those who will inherit salvation." This verse tells us the angels were created for a specific purpose that would be accomplished through serving God and man in the present history of the world, not one that happened before Genesis 1. Let's look at the evidence.

First, Job 38:4-7 tells us that a host of angels were present with God when the "foundations of the earth" were laid. The "foundations" must be referring to the establishment of the natural laws, such as gravity, in day 1 and the making of the dry land and seas in the third day. We are also told that the angels operate in the sphere of the universe and dwell in the heavens, where they are often referred to as "stars." The context of Revelation 12:4 makes it clear that the stars spoken of there are also the fallen angels spoken of in verse 9. Lucifer is called, "star of the morning" in Isaiah 14:12 and we are told in Ezekiel 28:13 that Lucifer was created on a "day" and in that same verse that he was in the garden of Eden.

The Ezekiel passage also suggests that Lucifer was created with the other Cherubim as their leader. Psalm 104 is a creation psalm and verse 3 makes it clear that the angels were made after the, "stretching out of the heavens like a curtain," in verse 2. This is an obvious reference to day 1 and day 2 of the creation week. Genesis 2:1 says, "Thus the heavens and the earth were completed and all their hosts." The angels are often referred to as the "host of heaven" or "heavenly hosts." Psalm 103: 20-21 is a passage that speaks to the angels and commands them to, "bless the Lord all you His hosts."

If God's creative work was done ultimately to meet the needs of man, then it would be consistent to understand that the angels were created before man so they would be in place to do their work of ministry for man. This would satisfy the principle established in Genesis 1 that God always makes the provision for a need before He creates the need. This principle is very obvious in Genesis 1. On the fifth and sixth days, God created animals and man possessing life. Before He created them though, He created all the things they would need to sustain their lives. He provided, water, light, atmosphere to breath, seas to swim in, air to fly in , and land to live on, vegetation to eat, and the sun, moon, and stars to give them a rhythm of life in relation to time. This provision principle strengthens our faith in times of need to know if God has allowed the need, He has already provided the answer to that need. The most profound example of this truth is salvation. Before Adam and Eve sinned, God already had a plan ready to redeem us through the Lamb that was slain before the foundation of the earth. Do you have a need? Ask God to show you His provision.

Angels created before man as ministering spirits to man in behalf of God also suggests that man was created in a position that was higher than the angels. Psalm 8:5 tells us in the New American Standard Bible that man was created a little lower than God. Hebrews 2:7 quotes Psalms 8:5 and translates the Hebrew word *elohim* as "angels" instead of God. The word *elohim* is translated "God" in Genesis 1 but is sometimes translated "angels" in other verses of scripture through out the Bible. It seems man was created a little lower than God originally but became lower than angels when he fell to a fallen angel's deception (Satan) in Genesis 3. The fact remains that ultimately man is a higher creation than angels. Consider the following observations concerning man and angels. Man was created in the image of God and God breathed into him the very essence of Himself (Genesis 2:7.).

He only did this for man, not angels. The angels were created all at once in all their hosts just like the animals and stars were created in hosts or large numbers. Man, on the other hand, was created as an individual person. God was willing to redeem man but not angels. The angels are ministering spirits to man on the behalf of God. The saved are going to rule and reign with Christ we are told in Revelation 20:4, not angels. Paul tells us that Christians will actually judge angels in I Corinthians 6:3. Man was given dominion over all the creation before the fall. Angels were never given dominion over anything. Satan is only the god of this world because man listened to his lies rather than trusting God, but Jesus has restored man's dominion by His victory over sin and death on the cross. No human being should ever have low self esteem. Low self esteem is a product of evolution's lie that man is just an animal not far removed from apes.

That same lie is a reason why many human beings act more like animals than the special creation they are in the sight of God. That will change when man starts learning about who we really are. Those who are created a little lower than God.

Naturalistic evolution does not know what to do for an explanation of the existence of angels let alone the spiritual realm. How could they have evolved? Besides, the natural universe is all there is supposed to be. And yet, there is a lot of phenomenon that takes place that cannot be explained by natural processes. Their response is to sweep it under the rug and pretend it is not there. Their hope is it will go away if they ignore it long enough. The fact is that awareness of the spiritual realm has grown rather than diminish in the last century.

OVERVIEW OF DAY ONE

Genesis 1:1 establishes the creation as a continuum made up of three interdependent ingredients, time, space, and matter. You cannot have one without the other. Time measures matter in space. Space with out time and matter is nothing; and matter must have space and automatically establishes time as the measurements between forms of matter. You cannot have one without the other. The continuum is irreducible. It is a perfect expression of the Trinity God Who created it. God, as a trinity, is a spiritual continuum. He is one God, just as the universe is one, and yet He expresses Himself as three individual persons, Father, Son, and Holy Spirit, just as the universe is understood as time, space, and matter. The universe continuum is a perfect expression of its Creator.

On day 1, we see God created the time, space, matter continuum that He began to stretch out in the midst of hyperspace (spiritual dimensions beyond the physical universe). The word "heaven" or *raqia* in the Hebrew actually means "stretched-out-thinness." Other passages tell us that the heavens are expanding. Isaiah 40:22 speaking of God says, "... Who stretches out the heavens like a curtain and spreads them out like a tent to dwell in." This is important because science tells us on the basis of the studies of red shift that the universe is expanding and the creation model actually predicts that would be the case.

The universe continuum started expanding as God then energized particles in the center of the universe by His Spirit. Up to this point in time the particles were "without form and void" or unenergized. We are told the Holy Spirit began to form the building blocks that would make up all matter. We see this energizing of particles into the first form of matter (water) when the scriptures tell us that the Spirit began to move, undulate, or energize those particles into atoms that formed into water. It is important to understand that energy cannot produce its self in a vacuum. It requires an energizer just as the creation account tells us. The Hebrew word *rachaph* for move or hover can also be translated "vibrated." This is interesting because all forms of energy come in the form of vibrations or waves. There are radio waves, tremor waves, light waves, micro-waves, etc. God Himself is the energizer of all matter. It is no wonder that Paul says in Colossians 1:17 about Jesus, who is the Creator in the flesh, "and He is before all things, and in Him all things hold together."

The water was formed in the midst of the darkness of the "deep" referring obviously to the space that was expanding before God brings light into the creation in verse 3. Later the "deep" refers to the waters of the deep. It is not mentioned but certainly understood that the water was formed into a sphere. How do we know it was understood to be a sphere? The book of Job is probably the oldest book in the Bible after the *toledah* in Genesis. It was probably written about the time of Abraham or before not too long after the flood of Noah's day. Job 26:10 is a creation account that rehearses the events in day one. The verse says, "He inscribed a circle on the surface of the waters, at the boundary of light and darkness." This must have been information passed on in extra Biblical literature that came from the pre-flood era.

The chronology of Genesis 10 helps us to see that Shem, Noah's son, lived at the same time as Abraham and could have passed that information on to men like Job. Solomon must have had the same information because he wrote the following in Proverbs 8:27 about wisdom, "When he established the heavens, I was there. When He inscribed a circle on the face of the deep..." So, the ancients understood that Genesis 1:2 refers to a sphere of water. They knew from the beginning that the world was round. This would have been a common sense deduction since all other bodies we see in space are round. God Himself could have told Adam and Eve this as well while they still had fellowship in the garden. This sphere of water must have been in rotation for we see that from the first day there was an evening and morning.

The fact that this sphere of water had an evening and morning tells us that the light would have been focused on one side of the rotating ball allowing for light and darkness to be separated on the surface of the earth as it is to this day.

And so, the earth, as a rotating ball of water, is the picture given to us of what took place on the first day. This is a major contradiction to the concept of a big bang that has the earth finally cooling into a small insignificant planet amongst other forming planets and stars in galaxies scattered across the universe. The first day suggests to us that the earth is the most significant planet in the universe and God's special focus of His creation work. It was not yet "good" though, because the first phase would not be completed until the first half of the third day. Only God's given light is truly "good" at this point. This light was the very expression of our good God's presence in His creation ready to provide what would be necessary to bring life to His work on days 5 and 6.

One final point needs to be made here. The first day having an evening and a morning makes it clear that the establishment of the measurement of time began from the very first day. The whole Biblical record from beginning to end is given in a historical context. It is a progressive revelation of God's redemptive work in the context of historical time. The scriptures being one long historical record is what ties together all the historical events found in them from beginning to end. Let me give you an example.

The historical creation account and the historical fall account establish the reason for the historical virgin birth of Jesus and His historical resurrection. It is the fact that they are all historical accounts in God's progressive revelation that links them all together into one meaningful on-going redemptive work of God in history.

The historical creation and fall established the need and the virgin birth and resurrection of God's promised Messiah provided God's answer to the need. My point is that redemption was all done in a historical context that ties it all together. That historical context began with the first day of creation.

God is sovereign over time, space, and matter. He is also sovereign over all the events of history in time including the second coming of Jesus Christ.

CHAPTER 4
WHICH ATMOSPHERE?

THE CONTROVERSY

What is described in day 2 has to be the most eye opening information we encounter in the creation account. What we learn there answers so many questions about what scripture has to say about so many other phenomena found in the rest of chapters 1 to 11.

Day 2 also separates the Biblical record from the concept of uniformitarianism more than any of the other five days of creation. On day 2, we are introduced to the water covering or canopy, as it is called by some. This protective covering would have made the environment on earth totally different from what it is now. \

It totally undermines the concept that "the present is the key to understanding the past," as uniformitarianism postulates. Instead, day 2 establishes that the past (the Biblical record) is the key to understanding the past and the present. If the water canopy theory is true, and I believe it is, then it changes everything that is now being taught in secular science about geology, the history of our atmosphere, paleontology (fossils), anthropology (man's history), "old age" dating methods (carbon dating and radio metric dating), and many other studies in science. It is no wonder that the water covering (canopy) theory is so widely resisted by scientists and theologians alike.

Let me offer a disclaimer at this point of our study. A workable model for showing how the water canopy can be explained scientifically has not yet been totally worked out at this writing.

There were so many variables involved in the make-up of the canopy that creation scientists, like Larry Vardiman of the Institute in Creation Research (ICR) in San Diego, California, are still working on the details. There are many questions that have to be answered by trial and error because there is little information available about the physical make-up of the canopy.

We do not know how much water it contained, although there had to be enough to allow for it to rain forty days and forty nights during the flood. You must remember though, that some of the rain would have been water that returned to the earth after having exploded into the atmosphere from the "fountains of the great deep," as we read of in Genesis 7:11. This added water would effect our calculations of the exact amount of water that made up the water canopy. We do not know what form that the water actually took, whether it was a liquid, ice, or vapor. We do not know how far it stood above the surface of the earth or how far it extended into space.

These as well as many other details have to be worked out before a workable model can be presented scientifically.

If all the above is true, then why do we believe there was a water canopy above our present atmosphere before the worldwide flood described in Genesis 7? There are several reasons.

First, the account of day 2 makes it clear that there was water above the atmosphere (firmament) and below the atmosphere. Second, Genesis 5 tells us men lived for hundreds of years before the flood. Genesis 11 then tells us human life spans shortened rapidly after the canopy was removed during the flood. Third, there had to be enough water above the earth to provide for the "gates of heaven" to be opened for forty days and forty nights as we are told in Genesis 7:12.

Fourth, the Biblical record insinuates in Genesis 8:22 that there were no seasons until after the flood, meaning there were year round even temperatures (we will see what this means later). Fifth, it also insinuates in Genesis 9:12-13 there were no rainbows until after the flood because there were no clouds (we will study this later). Sixth, the fossil record has several indicators that the environment before the flood was very different from the way it is now. A water canopy explains those differences very well (we will study this later). Seventh, other planets like Venus, Jupiter, Saturn and some planet moons have gaseous canopies suggesting a canopy was not unique to the earth. Eighth, in Joseph Dillow's book, *The Water Above*, he relates several flood stories in antiquity that suggest a covering was removed by the flood.[11]

These are eight circumstantial evidences that support the water canopy theory. Even if we have not been able to show how the water canopy worked scientifically, there is more than enough evidence in the Bible, in the fossil record, and from archeology to give us confidence that the canopy did exist. We need to remember that it is possible to get a verdict in a court of law on circumstantial evidence alone.

If the Bible is our final authority, than we need to trust what it says and let it lead us into the truth. Over the centuries, man's science has been shown to be wrong and God's word proven right. I believe this is one of those Biblical realities that will finally be verified by science.

THE WATER CANOPY

Genesis 1:6-8 tells us what took place on the second day of creation. It says that God took some of the water that made up the newly created ball of water and lifted it above the remaining ball of water.

In the process of lifting the water above the water, God energized particles into new elements to form the gases that would make up the earth's atmosphere. The result of this new masterful work of molding new elements into gaseous matter was a ball of water incased in an atmosphere of new gases (78% nitrogen, 21% oxygen, and 1% of several other trace elements) that were then incased in another covering of water. This is the description that makes the most sense when we read Genesis 1:6-7 that says, "Then God said, 'Let there be an expanse (atmosphere) in the midst of the waters, and let it separate the waters from the waters.' And God made the expanse, and separated the waters which were below the expanse from the waters which were above the expanse, and it was so."

It seems that God formed the expanse (atmosphere) in the midst of the water and, as the expanse formed, it lifted the water it was under out of the original ball of water and held it above that original ball of water.

This lifting up of the water would make a lot of sense if the water being lifted up was transformed into water vapor because we know that water vapor gas is lighter than the gases in our atmosphere. Couple this with the establishment of the electromagnetic field and the ozone layer (40 miles above the surface of the earth), and there were enough forces in place to hold a vapor canopy above the atmosphere. So you have three concentric balls incased within each other. The ball of water that would become the earth was incased in a ball of atmosphere the Bible calls the expanse. The atmosphere was then incased in a ball of water that was most likely in vapor form.

Let us consider a little further the form (liquid, ice, vapor) the water was in that made up the canopy above the earth. I have already suggested the canopy was made up of water in vapor form but others have tried to show that it was either in liquid or ice form. Those who argue that it was in liquid form do so mainly because the Biblical record says it was water and does not directly say it was ice or vapor. There are several problems with it being liquid water. First, there is not a physical mechanism in place that can explain how that much water was held above the earth. Either the mechanism disappeared after the flood or the water was held in place by God's miraculous intervention that overcame the present laws of physics.

Normally, God relies on His created natural laws. Miracles are the exception not the rule in nature's normal functions. Second, the water would have to be totally clear and stable in order for the heavenly bodies to be seen from the earth's surface.

The surface tension between the water and the atmosphere would cause the water to convert to droplets that would develop a clouded look that would hamper visibility. Third, as the water ascended into the upper levels of the atmosphere and out into space, it would either turn into ice or vapor. There is a heat sink eleven miles up where the water would freeze if it collected there. It would vaporize if it went farther out as it encountered the heat of the sun. Fourth, when the fountains of the great deep exploded into the canopy it caused a torrential rain of water. If the canopy was already in liquid form, what kept it from falling before the flood as rain? There is no valid answer to that question.

The canopy being in the form of ice seems to be more feasible than liquid but not as explainable as vapor. Dr. Carl Baugh in his book, *Panorama of Creation*, gives an explanation of how an ice canopy might have worked.

There is a heat sink 11 miles above the earth that maintains temperatures of -130 to -180 degrees Fahrenheit. Dr. Baugh speculates a thin ice canopy could have been maintained there.[12] Joseph Dillow in his book, *The Water Above*, gives a lengthy explanation of all the problems one faces in trying to show how an ice canopy would work, given the present laws of physics that would have to be overcome.[13] Although ice is a possibility, vapor is the most plausible form the water in the canopy above the earth would have taken. It would have been clear, lighter than the atmosphere and held in place by present laws of physics. This being the case, we will assume that the canopy was in vapor form for the purposes of our study and the development of our creation model.

Regardless of what form the water took in the canopy above the earth, there would have been two major impacts this covering of the earth would have. First, the mass of the canopy being pulled down on by the gravitational pull of the earth would have increased the atmospheric pressure on the surface of the earth. It has been speculated that the atmospheric pressure was as much as double what it is now. Second, the light coming from the sun would have been filtered in a way that the long waves of light would still pass through, whereas the dangerous short waves (ultra violet radiation, cosmic rays, etc.) that cause mutations would have been reflected away from the earth.

This heavier atmospheric pressure and the reflected short light waves would have created an environment far different from the post flood environment we now live in. Using these changes in environment, we can explain the phenomena in the Genesis record that many use as reasons why they accept it as mythology rather than literal history.

They also explain differences we find in plants and animals found in the fossil record. We will look at each of these two environmental characteristics to show the differences they would have made that no longer exist.

REFLECTED LIGHT

We have already talked about the two different models based on either evolution or literal creation. This reflected light wave characteristic of the vapor canopy theory provides the foundation for several predictions that are supported by the Biblical record and the fossil record.

For one, the canopy would be heated by the light waves and would cause the heat in the canopy from the light to be evenly distributed over the whole surface of the earth. The surface of the earth would basically be a world wide greenhouse. This even temperature on the surface of the earth would have several impacts. It would mean that there would be no winds or violent storms that we have today. There would be calm breezes as temperature moved between 72 and 78 degrees.

This would also keep the atmosphere clear of dust not allowing the formation of clouds and thus not allowing the process for rain. This makes Genesis 2:5 understandable when it says, "... The Lord God had not sent rain upon the earth." Without rain and clouds in the atmosphere, it also causes the establishment of the rainbow in Genesis 9:13 after the flood to have a great deal of meaning suggesting there where no rainbows before the flood. The even temperatures would also provide a year round stable environment that would keep plants, animals, and man from having to continually adjust to broad changes in the weather. Year round even temperatures would remove the reason for several diseases and would slow down the aging process.

It would then be possible for man, animals, and plants to live much longer, as the Bible suggests in Genesis 5.

Even temperatures also mean that there were no major changes in the seasons and thus allowing for a year round growing season for vegetation. Planting and harvesting could take place year round. The growth rate of world wide vegetation would be phenomenal. It is no wonder the coal seams and oil deposits in the earth are so abundant after that original earth was destroyed in the flood.

Amos 9:13 is a prophecy of the millennium after the second coming of Christ, when several scholars believe the pre-flood world environment will be restored in the "regeneration" Jesus spoke of in Matthew 19:28. The verse in Amos speaks of year round planting and harvesting when it says, "Behold, days are coming, declares the Lord, when the plowman will over take the reaper and the treader of grapes him who sows seed..." The picture is one of a man planting in one field while another man is harvesting in a different field. That is what it would have been like before the flood with even year round temperatures.

The global even temperatures, including the poles gives grounds for several predictions that the creation model can make. For one, the creation model predicts that under the polar ice caps and in the permafrost of the far northern regions, remains of warm weather plants and trees would be found. That is exactly what is being found. In the Antarctic, evolutionist explorers are now being forced to admit that that region once had a much milder climate that produced lush vegetation. In the northern regions of Canada and Russia as the permafrost melts away they are finding the remains of fruit trees and graze lands. This includes finding large amounts of flowers in the stomachs of frozen mammoths.

There are the remains of warm weather animals also being found in those regions. All of this evidence suggests that there was once a much different environment on the earth than what exists now. The canopy before the flood is a very reasonable explanation for this different environment. The post flood world, before the ice age, adds to the explanation.

HEAVIER ATMOSPHERIC PRESSURE

A heavier atmospheric pressure would have had several positive influences on the pre-flood environment as well. We see the results of these influences in the Biblical record, the fossil record, and the advances being made in science today. We have already seen that the water above the earth, even in vapor form, had mass that would put pressure on the atmosphere it was above as gravity pulled the watery mass toward the earth. The gases of the atmosphere would be compressed by the weight more than it is at this time when the canopy is no longer there.

The Biblical record tells us that Adam lived 930 years and Methuselah lived 969 years. All the descendants of Adam named in Genesis 5 lived several hundred years. We have learned from science that our physical bodies actually operate more effectively under heavier atmospheric pressure. Hospitals are now using hyper-baric chambers to increase the healing process in our bodies. They are effective in healing sores and wounds of diabetics. In a heavier atmospheric pressure, our bodies would carry more oxygen to the cells of our bodies through our red blood cells, than we do today.

The heavier pressure would also increase the ability of more blood being allowed to get to all parts of our bodies more efficiently with more oxygen being absorbed by the cells.

This would definitely have a huge impact on our body's ability to heal itself and absorb more nutrition, thus allowing it to stay healthy much longer.

There are accounts of people in deep sea diving capsules who have experienced rapid healing of open wounds while under the heavier atmospheric pressure.

Speaking of nutrition, the plants before the flood would also be a more pure source of nutrition due to the fact that no mutative influences would be coming from the light source and they would also have an abundant source of nutrients to draw from the ground and absorb those nutrients into their cells more effectively. This would make the plants a much richer source of nutrition. They would also grow to be much larger, as we have discovered in the fossil record. Small ferns have been found that were as large as trees at one time. There are many examples of larger plant life in the fossil record.

The fossil record also makes it clear that animals were once much larger than they are now. There have been remains of beavers unearthed that were the size of a man. There were sloth that were larger than large bears. Insects where much larger than they are now. It is now being suggested that the large dinosaurs grew so large, because they lived for so long. They also tell us that the atmospheric pressure had to be much greater to help their large bodies absorb enough oxygen in order for them to survive. There have been dinosaurs found that were one hundred feet long and weighed as much as one hundred tons.

The question today, given our present atmospheric pressure, is how could they have breathed? The great blue whale can help us answer that question.

A blue whale is the largest mammal on earth at one hundred feet long and weighing in at one hundred tons.

It breaths with lungs just as the large dinosaurs did. One reason why it is able to breath and absorb the oxygen into all the extremities of its large body is the fact that when it dives deep into the ocean the atmospheric pressure increases and helps the huge whale's cells absorb all the oxygen they need. Land animals today do not get any larger than elephants because the present atmospheric pressure would not support an animal much larger than an elephant.

The fossil record also tells us that there were large reptilian animals that could fly with wing spans that reached fifty feet in length. These animals were as large as fighter jets we have today. Even with hollow bones that were lighter than the land animals, these animals would have weighed several hundred pounds. The big question is how did they get off the ground to fly if the atmosphere was as thin in their time as it is now? The answer is with great difficulty. The largest birds of our day are the albatross often referred to as gooney birds because of how they often fall flat on their beaks as they run down the beach trying to get airborne. It is all they can do to get off the ground. If these birds, not nearly as large as those flying reptiles, have a hard time getting air borne in the present atmospheric pressure, so much more the problem for flying reptiles. The canopy tells us that those extremely large flying reptiles could fly effortlessly if the atmospheric pressure was twice what it is now.

The vapor canopy helps explain several phenomena in the Bible and the fossil record that evolution has to try to explain away or just over look.

Even though a physical model has not been fully developed as of yet that shows how the canopy existed, we have more than enough circumstantial evidence in scripture, the fossil record, and stories from ancient cultures to support our faith that it did exist before the flood of Noah's day. Obviously it is a faith issue, but it is a faith rooted and grounded in objective reality. So, when it comes to accepting the revelation of the second day, we do so by faith that is supported by science and other disciplines. The reality is that if the second day is literally true then evolution and old earth arguments become meaningless. A worldwide flood, on the other hand, becomes more than plausible because the one (canopy) makes the other (flood) possible.

THE REGENERATION

I mentioned earlier the regeneration Jesus spoke of in Matthew 19:28 where Jesus told His disciples, "...Truly I say to you, that you who have followed Me, in the regeneration when the Son of Man will sit on His glorious throne, you also shall sit with me upon twelve thrones, judging the twelve tribes of Israel." This is a reference to Revelation 20:4 that says, "And I saw thrones, and they sat upon them, and judgment was given to them."

Revelation 20 is the chapter that refers to the thousand year (millennium) reign of Christ on the earth. It is important to note that Jesus uses the word "regeneration" to refer to His millennial reign. This is important because there are several passages in the Old Testament that suggest the earth will be restored to its pre-flood environment that will include the water vapor canopy.

Some scholars suggest that all the physical upheaval during the tribulation period prior to Christ's second coming describes the process the earth will go through to be restored to that pre-flood environment.

Amos 9:13, "... the plowman will overtake the reaper..." has already been shown as a millennium prophecy that tells us there will once again be year round planting and harvesting rather than the seasons we now have. Other passages tell us animals will not be carnivorous again. Isaiah 11:6, "...the wolf will dwell with the lamb..." and Isaiah 65:25, "...the lion will eat straw like the ox..." to mention two. Apparently animals did not eat each other until after Genesis 9:1-5.

We will study this later. We are also told people will live a long time once again in Isaiah 65:20, "... for the youth will die at the age of one hundred and one who does not reach the age of one hundred shall be thought accursed." Under the canopy Adam lived 930 years. To people who lived that long one hundred years would be equivalent to 10 years in our present life spans of 70 to 100 years.

Another phenomenon that will exist in the millennium is there will not be any poisonous animals again. Isaiah 11:8-9 tells us, "...and the nursing child shall play by the hole of the cobra and the weaned child will put his hand on the vipers den. They will not hurt or destroy in My holy mountain..." These verses tell us there will be a regeneration of the electromagnetic field back to its original strength. Poisons are made up of proteins that become toxic when they bond together. Higher electromagnetism does not allow those proteins to bond.

Dr. Carl Baugh of Glen Rose, Texas Creation Museum has done studies on snake venoms that show this to be true.

The earth's environment before the flood provided an existence we can only dream about in our day in time. It is just like the Creator to return to rule on the earth He created in the environment He intended before He had to bring His judgment against it with the flood. This regeneration once again affirms God's word is consistent with itself. How we found the earth to be in the beginning will be how we will find it at the end.

THE COVERING PRINCIPLE

The canopy of water protecting the earth establishes a principle that is found throughout scripture. The principle is what I call the covering principle. The principle is God always provides a covering for those who live under His protective grace. God gives five specific coverings in scripture. Of course the water canopy in Genesis 1 is the first covering.

Second, after Adam and Eve sinned, God provided the skins of animals to cover their nakedness in the hostile environment outside of the Garden of Eden (Genesis 3:21).

Third, the ark was a covering given by God to protect Noah, his family, and the animals from the flood (Genesis 6:14). The word "pitch" in which the ark's wood was soaked can also be translated covering.

Fourth, when God brought Israel into the Promised Land, He gave them the Day of Atonement (covering) to protect them from His holy presence in the land.

Fifth, in I Peter 3:18-21, Peter compares Jesus to Noah's ark. Just as God brought those in the ark safely through the judgment waters of the flood, Jesus is the ark of God Whom God will bring believers safely through the final fiery judgment. Those who have put their faith in Christ for their salvation are covered in Him. I Peter 3:21 speaks of a baptism.

Baptism means an emersion into something. Peter's point there is that just as those in the ark were carried to safety in the flood; those baptized into (covered by) Christ positionally by faith will be carried to safety in the final firey judgment as well. Jesus Christ is our ultimate covering.

The covering principle is all about living in the covering God has provided for those who choose to obey Him. The covering is provided by God for those who choose to live under His grace.

CHAPTER 5
WHICH
ENVIRONMENT?

We now come to the third day of creation. The text tells us that day 3 involved two separate pronouncements of "good" suggesting God finished two more levels of creative activity on this day. First there was the creation of the land and seas, and then there was the creation of vegetation. This vegetation would grow out of the land providing a source for food and a mechanism for purifying and replenishing the atmosphere.

This is a far cry from the evolutionistic view that suggests that original particles continued to condense into a smaller and smaller compact ball that finally exploded creating an ever expanding space of gases collapsing into heavier elements that finally produced burning stars in rotating galaxies while other elements over millions of years finally cooled into planets orbiting those stars.

The earth, of course, would have been one of those billions of cooling rocks that just so happened to develop an atmosphere and water. This specially endowed rock became the necessary environment that could support simple life forms that would develop the capacity to evolve into more and more complex forms of living organisms.

And so, here we are once again faced with two views that are complete opposites suggesting that one is true or the other is true but they both cannot be true. They are contradictions to one another.

This reminds us of an important issue of today. Is truth absolute or is truth relative to each individual's personal preference or a society's interpretation of truth in their particular place in time? One example would be the two ways of interpreting the United States Constitution in our courts these days. These two views are at odds with one another being made evident by the selection of federal judges by the President and approval by the Senate majority.

Some judges are strict constructionists and others are activists. The constructionists believe the Constitution is a document intended by the founding fathers to be static and unchanging from their original intent. In it, they established principles that would always be the same and be the guidelines for the making of new laws based on those unchanging principles. Activist judges, in contrast, believe the Constitution is a living document to be reinterpreted according to the values of the present generation not according to the founding father's original intent or values.

Thus, we now have activist judges in federal courts who rewrite the Constitution in their decisions based on present sentiments, not the original principles and intent found in the Constitution. How can this happen? Our founding fathers were guided by their Judeo-Christian worldview. Many judges these days, trained in liberal law schools, have a relativistic worldview that sees truth as relative to the present age and not absolute.

The philosophy behind this worldview is naturalistic evolution that reasons, if all reality is a product of random chance and original chaos, then there is no God and there are no absolutes, especially in the realms of morality and law. The logical conclusions of evolution theory take us to an illogical philosophy that truth is what you want it to be. In other words, when it is all said and done, nothing is really true.

Everything is situational and relative only to what is accepted at the present time by those who control the powers of indoctrination. But, if the Bible account of creation is true, then there are absolutes that were established by a moral, unchanging Creator. These absolutes are always true regardless of the situation.

FORMATION OF LAND AND SEAS

In this chapter, we will note God's creative activity on the third day and give evidences that support the creation record. The first part of the third day tells us that God produced new elements out of the sphere of water that were energized to form the matter that would make up the core, mantle, and crust of the earth. In these liquid, gas, and solid forms of matter were contained all the necessary elements for supporting living organisms that would be formed on the fifth and sixth days out of the same elements.

It is amazing that just in the last couple of hundred years we have learned through scientific investigation that our physical bodies have the very same elements as rocks and dirt made up of different combinations of atoms. The Bible has been telling us that fact from the very beginning.

As we look at evidence for the creation in the rocks and soils of the earth, we need to remember that, what was a result of God's original creative activity on the third day, would have been totally rearranged by the year long cataclysm of the flood in Noah's day. Much of the water under the crust of the earth called " the fountains of the great deep" would be in the oceans now. The original crust of granite rock would be broken up now into floating plates due to the break up of the crust in the flood year.

This breaking up of the crust would include the sinking of much of the crust under the oceans while other parts were thrust up to establish the new continents and mountain ranges we see today. Also, a great deal of melted magma from the core of the earth is now on the surface of the earth and in the oceans due to the great amount of volcanic activity that began during the flood year and has continued to this day. There should be indicators, though, that these new formations of elements came into existence recently, not over billions of years.

One such indicator is the earth's electromagnetic field. Dr. Thomas G. Barnes did a study several years ago on the measurements of the decay rate of the electromagnetic field and discovered that it has a half life of 1400 years. That means if you start moving back in history taking measurements of the magnetic field every 1400 years you would find that it was twice as strong as it was 1400 years later. Go back more than 10,000 years and the magnetic field would be stronger than that of a magnetic star. That means that most of the physical processes necessary to sustain life on earth would not be able to function.

What does that tell us? It tells us there was a time in history less than 10,000 years ago when the electromagnetic field of the earth was operating at an optimum level that has since begun to deteriorate under the second law of thermodynamics or slow dying of all processes. Dr. Barnes' study suggests the earth is less than 10,000 years old.[14]

It is theorized that the electromagnetic field is produced by the rotation of the core in the center of the earth. If that is true, than the field would have been a result of the transformation of the waters into the core, mantle, and crust of the earth on the first half of the third day.

The magnetic field would have been operating at the perfect level of strength that provided the necessary environment for the soon to come chemical reactions in plants and living organisms. An interesting note is, if the core is rotating inside the earth now and that rotation has been slowing down at a predictable rate in relationship to the deterioration of the magnetic field, then the earth should have melted 20,000 years ago because of a much faster rotation rate than now; that is, if the concept of uniformitarianism is true.

Other studies show that the electromagnetism has reversed itself in the past possibly during the flood and the after math. A post flood ice age and continental drift may have also caused reversals that weakened the electromagnetic field.

Another indicator that supports the creation story that the earth was created rapidly just a few thousand years ago has been suggested by Robert Gentry as a result of his studies of polonium radiohalos. Polonium is theoretically a daughter element of uranium as it decays.

Polonium 218 has a half life of 3 minutes that leaves concentric balls that look like halos when cut in half. These radiohalos have been found in mica and fluorite without any evidence of uranium having been present, suggesting that the polonium was present at the moment of the creation of these rocks. This would be the case if all the elements were created at the same time, even though some now deteriorate into daughter elements because of the fall and curse. If there was no decay before the fall, then all the elements would have been needed to be created to sustain all the necessary chemical reactions needed for optimum existence. If the evolution theory of the formation of the crust of the earth requiring millions of years of cooling is true, the radiohalos would never have survived.

The fact that they exist suggests these basement rocks of granite formed very rapidly with no cooling time.[15]

Some creation scientists have questioned Gentry's conclusion that the polonium radiohalos are a result of the original creation but can be shown to be a natural occurrence from the processes involved in the year of the flood. The ICR RATE team did an extensive study on these radiohalos. They discovered that many of the halos seeming to be parentless actually were the result of transference from the parent element uranium to their present place of decay.

The flood and the tremendous amount of upheaval and transfer of elements explain the isolated radiohalos very well. It was their study of radiohalos that gave strong evidence that decay rates were greatly accelerated during the flood. One thing that is clear about polonium radiohalos is they in no way can be explained by evolution theory.[16]

One other indicator that the earth was created recently has to do with the huge amount of helium still found in the earth. There have been studies done by the Institute in Creation Research's RATE program (studies done on the age of the earth) that show how zircons found in granite contains high levels of helium that has not been defused like it should have if the earth is billions of years old.

Over time, helium has the ability to make its way out of even the hardest of substances, such as zircons. It would finally escape from these hard crystals in a few thousand years, not millions of years. Whether the helium became trapped in the zircons as the core and mantle of the earth formed on day three or was produced during the volcanic activity of the flood and after, the point remains that this phenomenon strongly suggest the earth is only a few thousand years old. The helium should not be there in such high amounts if the earth is old.[17]

These three phenomenon: electromagnetic decay, polonium radiohalos, and helium still trapped in zircon crystals, are predicted by the creation model. On the other hand, they have to be explained by the evolution model because they do not fit their old earth assumptions. There are many other studies in geology and physics that give similar evidence to the young earth assumptions of literal creationists. We will look at those in more detail when we get to our study of the flood.

Not only was the land formed on the first part of day three, the seas were also gathered into one place. It is not quite clear as to what the meaning of gathering the waters into "one place" actually means. We do know that the gathering of the waters was called seas.

Because of the total transformation of the topography of the earth due to the violent forces brought on by the flood, we must understand that what is described here in the original creation was totally different than what we now see after the flood.

There are two formations the Bible tells us the water that was not formed into dry land was gathered to. On the surface of the earth, there were seas or large bodies of water similar to large lakes but not necessarily what we now know as oceans. The other place, where we find a body of water, is called the "fountains of the great deep" in Genesis 7:11.

Apparently, when the mantle and crust of the earth were formed, a large amount of water was trapped under the surface of the earth. This water under the crust of the earth would have been under heavy pressure due to the weight of the earth's crust and the heat coming from the core. There must have been underground passage ways that connected underground reservoirs around the earth that also had outlets for the pressurized water to escape to the earth's surface. These outlets produced large springs of water that were the source of huge rivers, such as the one found in the Garden of Eden that divided into four other rivers beyond the garden. Those rivers must have emptied into the seas that allowed the water to return to the "fountains of the great deep."

This water cycle would produce a world wide watering system for the whole earth. A circulating fountain system of sorts. With this kind of watering system, rain would not have been necessary. There was plenty of water in the earth itself to allow for a mist to rise on the surface of the ground as we are told in Genesis 2:6. This kind of system would have provided a semi-hydroponic type environment ready to support the abundance of vegetation God was about to create on the rest of day three.

Some creation theorists believe that the seas were one large body of water, and the land was one large body of land (the pangea theory) not broken up into continents until after the flood. There are indications that the continents were once inner connected. Whether the seas were separate or made up one large body of water, the inference is that animals and man would have been able to spread all over the earth equally. There were no natural barriers, like oceans or uncross-able mountain ranges, to hinder worldwide migration. Given the fact that the canopy tells us even temperatures were world wide and different languages were not introduced until after the flood, man and animals would have evenly spread over the earth and humans would have maintained fairly stable physical characteristics worldwide rather than the diversity we see today.

We see this to be true in the fossil record when we find animal remains buried in places that are no longer their habitat after the flood. Rhinoceros and Hippopotamus that now live only in Africa have been found in the northern regions, to name a couple. At the completion of the forming of the land with all its nutrients and the bodies of water ready to sustain the plants to be formed next, God pronounced "good" the completion of this level of forming elements into new forms of matter.

FORMATION OF VEGETATION

Now we come to the remarkable work of God that will produce systems that have the capacity to transform the nutrients in the ground into carbohydrates with the help of the light of the sun energizing the process of photosynthesis within plants.

Not only making the plants usable to the coming animal and human life forms for food, but also giving the plants the ability to clean the air of carbon dioxide and transforming it into glucose and oxygen.

This would insure the quality of the breath of life that was soon to be created as a part of the new life forms on the fifth and sixth days. It is an amazing thing to think of how living things depend on the plants for oxygen production and the plants depend on living things for carbon dioxide production.

It is also amazing to consider the vast variety of plants there are both on land and in the seas. There are plants that are microscopic and plants that tower above the forest floor in the redwood forest. There are all kinds of plants of great beauty and others that are ugly and bothersome. Many plants can be used for food, some are poisonous, others can be used for medicine, and some even eat insects, such as the pitcher plant, sundew, and Venus fly trap.

The Biblical record tells us in Genesis 3:18 that some plants were transformed into thorns and thistles, suggesting these unusual and bothersome plants were not a part of the original creation but were transformed by God at the fall. The interesting thing about all the vegetation on earth is botanists have found very few if any plants in the fossil record that are different from what we see today. There are very few fossil plants that could be considered transitional forms and all of those are questionable. The kingdom of plants does not give evidence of evolution as some die-hard evolutionists would like for us to believe.

As a matter of fact, plants are an embarrassment to evolution theory. Dr. E. J. H. Corner, a respected evolution botanist once wrote, "... I still think to the unprejudiced, the fossil record of plants is in favor of special creation..."[18] Of course this is exactly what you would predict to be the case when the Bible tells us that God formed each of the plants according to their own kind with seeds in them that would only reproduce the same kind of plant the seed came from.

This third day of creation tells us clearly that God created each plant according to its own kind. They each have their own special seed that reproduces the same plant it came from. We will see later that plants do not posses life like animals and man but their cell structure is made up of DNA just as animal and human cells are.

We know that DNA is a strand of amino acids that posses codes that pass on information in the cells of the plants. That DNA code is specific for each plant and that is what makes them reproduce according to their own kind. A good friend, Dr. Norbert Smith, made an interesting point to me about that information in plant and animal DNA. Hebrews 12:2 tells us God is the author and finisher of our faith but the third, fifth, and sixth days where God created the plants, animals and man tell us God is also the author, information provider, of our physical body's total make up.

Think of it. As long as God knows the DNA code in each person, animal or plant, He can reproduce that physical body perfectly even if it is totally destroyed. God's word tells us every human body will be perfectly resurrected at the end of time. God can do it because He has the information He needs in every strand of DNA.

That brings up an important point. The creation account tells us that God formed the plants from the existing elements according to their own specific kind. Obviously these plants were also given the ability to adapt to different environments later when the environment of the earth changed after the flood and the loss of the canopy protection. But we also know that plants are not mobile and so they are not able to cross pollinate on their own. Not being able to cross pollinate would lead to an ever weakening plant life. They would need insects, birds and some animals to do pollination for them, especially if winds were not present because of the even temperatures.

Breezes would help but not like the winds do today. My point is that the plants were formed on the third day. Birds, insects, and other animals were formed on the fifth and sixth days. If the "old earth" model is true, than plants would have to wait thousands of years before they could cross pollinate after animals started to evolve enough to transport pollen. If they evolved together then the creation story is false and should not be accepted as truth. This is another one of those situations where one is true or the other is true but they both cannot be true. If the "old earth" interpretation is true, then the creation story in this matter of cross pollination does not even make for good allegory or myth.

The creation story left alone as it is makes perfect sense if you believe, "In the beginning God created the heavens and the earth." The only real difference in plants today and those in the fossil record is size. Plants in the fossil record are much larger than what they are today. Small ferns today were as large as trees back then. Mosses today that grow a few inches high are two or three feet tall in the fossil record.

Horsetail reeds were once as much as fifty feet tall rather than the five or six feet heights they grow to today.

That tells us there was a time when there were more nutrients in the ground, the water supply was more abundant, there was less negative light, and the atmosphere was much heavier than now. In other words the plants tell us from the fossil record that the environment of the earth was very different before the flood than it is now. The fossil record tells us another important truth about God's creative activity dealing with vegetation. Tropical plants, redwood trees, fruit trees, flowers and grasses have been found buried in the southern and northern polar ice-caps and in the permafrost of the northern regions of Canada and Russia.

Frozen Mammoths, as has been mentioned, have been found with flowers and grass undigested in their stomachs. I read a Strange But True article once that said, "270 million years ago, the continent of Antarctica was mostly covered with forest and marsh." As a creationist, I did not see it as strange and knew it was true because the creation model predicts that forests in Antarctica, as well as warm weather plants, is what you should find there because of the pre-flood environment. I had also already read that redwood forests had been discovered in Antarctica after scientists drilled core samples deep into the ice covered land.

That tells us that when God created, He covered the whole earth with vegetation to provide plenty of food for the creatures He would create on the fifth and sixth days.

Thus, the earth was covered with all the present varieties of vegetation on land and in the seas plus some extinct varieties. We have to use the word "extinct" loosely. The wollemi pine once thought to be extinct for thousands of years has been discovered in Australia. Its discovery was a real blow to evolutionist's "old earth" theory. Another tree to consider is the bristlecone pine trees found in the White Mountains between California and Nevada.

These trees are the oldest living things on earth. They are dated at about 5,000 years old. Why are they no older than 5,000 years if the earth is billions of years old? When you factor in the flood, that happened about 5,000 years ago, you realize that nothing on earth could be older than the time after the flood. This is another creation model prediction.

The third day came to an end and God pronounced for the second time that day that all He had done was "good." Now the only things left to form from the elements would be the holders of light in the space beyond the water canopy to project the ongoing light needed to provide the energy for the plants to produce food for the coming life forms and provide the energy for the life that those creatures would need as well as man. That also brings up another point. The plants could wait a day for God to transfer His light to the light holders He would make on the fourth day. They would not have to wait millions of years, but instead only a day. God is following His preplanned progression and it is going perfectly.

Obviously, God was working out a plan that clearly had nothing to do with any description fallen man would invent to try to explain how this universe could come into being by its own design.

CHAPTER 6
WHICH COSMOLOGY?

Now we come to the fourth day of creation. The fourth day is another of those parts of the creative sequence that totally contradicts the old universe, big bang, uniformitarian model. The big bang begins with particles that collapsed upon themselves to such a minute degree as to explode in such a way as to cause space to expand and have all those particles begin to form into necessary elements to form matter that would make up burning stars and cooling planets.

That is a far cry from God beginning with forming the earth in an expanding space out of particles, then atoms, and then elements into all forms of matter until the fourth day, when He then forms the sun, moon, and finally the stars. It is a total reverse order of progression. That is why it is so amazing to me that theistic evolutionists and progressive creationists still try to make it sound as though the two concepts are compatible. Even when you make the creation account an allegory, there is no way the two models can be made to resemble each other. They are two different children born of two different ideological parents that are not related. One is true or the other is true but they both cannot be true. They are opposites.

THE CLOCK

If you were stranded in the Sahara desert with no other human beings to be found for miles around and you came upon a Rolex watch lying in the sand, what would your first reaction be to finding the watch?

Why you would say to yourself in complete amazement, "Imagine that. Over millions of years the winds have blown and rains have come and over time gears, and springs, and small screws formed in the sand along with a casing, a crystal lens, a watch band, and when all the parts were ready they just happened to fall into perfect order to form this watch totally by random chance." Of course, you are thinking, "that is ridiculous," I hope. Everyone knows watches do not just happen by random events. And yet, that is exactly what many would have us to believe about how the sun, moon, planets, stars, and galaxies formed into the most precise clock known to man. The heavenly bodies we observe in the sky at night are organized in relationship with one another to form one big clock.

God tells us in Genesis 1:14, "...Let there be lights in the expanse of the heavens to separate the day from the night, and let them be for signs, and for seasons, and for days and years." This verse makes it clear that the sun, moon, and stars were created to give man a way to measure time as it relates to the progression of time from the beginning onward. That tells us that God, from the beginning, wanted man to be aware of where he is in time according to the beginning of creation. God's word is inspired by the Holy Spirit and apparently God wanted to make sure that the progress of time was recorded from the beginning. We see in chapter 5 that this whole chapter was dedicated to recording the life span of each of the patriarchs from Adam (930 years) all the way through Noah (950 years). And then, in chapter 11, the ages of those men descended from Shem are given all the way up to Terah (205 years). In the rest of Genesis, Abraham's (175 years), Isaac's (180 years), Jacob's (147 years), and Joseph's (110 years) ages are all recorded.

And so, we have the time frame from Abraham up to the time of Israel going into Egypt until the death of Joseph, who was 110 years old when he died. It is important to note that the last verse in Genesis or "generations" gives the age of the patriarch, Joseph. This all tells us that the original followers of God were very concerned to keep a record of the progression of time, apparently at the command of God that is suggested in Genesis 1:14.

If those men in Genesis were concerned about the progress of time, we should be concerned as well. It certainly is clear that God, in His word, was concerned about giving us a time frame in which He wanted us to understand His progressive revelation of Himself bringing about His redemption. It is apparent in scripture that God is working out His redemptive plan according to a time table in order that those who believe will not be caught unawares of God's ongoing plan. We see that Paul uses the phrase, "in the fullness of time" or "fullness of the Gentiles" to emphasize God's plan is being worked out according to a time table.

Daniel 9: 24-27 actually gives the time frame of when the Messiah would come after the return of the exiles from Babylon. The book of Revelation is written according to a time frame of seven years of tribulation. The scriptures make it clear that we need to take the time frame set out in those scriptures as grounded in God's intended prophetic nature in His word. We need to be aware that if God created in six days and rested on the seventh day and His word tells us that a day is as a thousand years to God, then according to the literal time frame of scripture, we are in the six thousandth year or sixth day, by God's standards.

This tells us that in the time frame of God's redemptive work, the second coming, the millennial (thousand years) reign of Christ, and the coming of the new heaven and earth is just around the corner. I am not saying that is what the Bible teaches. I am saying that on the basis of all the information about time in God's word, there are strong indicators that support the possibility of Christ's second coming after six thousand years of earth history.

We are told in Genesis 1:14 that the sun, moon, and stars would be used to identify signs and seasons. This being the case, it does not take much speculation to infer that Adam was intelligent enough from the moment of his creation to be able to read the stars in their constellations. The fact that we are told in Genesis 5:5 that Adam lived 930 years tells us he began counting his age from the time of his creation. If God created the heavenly bodies so man could know their signs and seasons, it makes sense that God Himself could have taught Adam and Eve how to read the stars before the fall. This would let them know at all times where they were in relationship to the earth's orbit around the sun and the moon's orbit around the earth, thus letting them know when each new season had begun and when each new year had begun. Yes, the universe is one big clock and God is keeping time.

MILLIONS OF LIGHT YEARS?

One of the major arguments "old-earthers" use to support their position that the universe is billions of years old is the fact that the stars that we can only see with telescopes are millions of light years away. This says to them that it has taken millions of years for the light from those stars to travel across the universe to finally be seen by us on earth and so the universe has to be very old.

Dr. Russell Humphreys, at one time a physicist at Sandia National Laboratories and now an affiliate of the Institute in Creation Research, has written an intriguing little book called, *Starlight and Time*. The book explains Dr. Humphreys' theory of how the light from the stars could be seen by Adam and Eve when they were created on the sixth day of creation without taking millions of years to reach the surface of the earth. He uses Einstein's theory of general relativity as the foundation of his theory.

I will try to explain his theory as briefly and simply as I can. Einstein's theory of general relativity basically or at least in part suggests that time is relative to its relationship with gravity. For example, a precise clock placed at sea level will keep time at a slightly slower rate than a clock placed at the top of a high mountain. The farther away from the center of gravity time moves the faster it moves in relationship to the time found at the center of gravity. According to Dr. Humphrey's theory, if the earth or at least its galaxy is at the center of the universe, as Genesis 1 infers, and the universe is limited by an outward boundary that has been expanding since the creation week, then the earth would be located close to the center of the gravitational pull of the universe. Dr. Humphreys then suggests that at the present rate of the expansion of the universe it would have been close to fifty times smaller than it is now. Applying the theory of general relativity to this model would allow the stars in their galaxies to be created closer to the earth's galaxy on the fourth day and then be expanded out from the center of the universe in concentric spheres.

But the light from those stars in their galaxies would move faster and faster in time the farther they moved away from the center of gravity of the universe where the earth is located.

This being true, time, measuring the speed of light millions of light years away from the earth, would be traveling at a rate millions of times faster than light moves in time at the surface of the earth. In other words, the farther light is from the earth the faster time moves in relation to time on earth. That would make it possible for all the lights in the universe to reach earth in a matter of hours or days. Adam and Eve would then be able to see distant stars and galaxies from the beginning. It is important to note that Dr. Humphreys has done further studies on red-shift and the cosmic microwave background (CMB). His studies have shown, along with other scientists, that earth's galaxy (the Milky Way) is at the center point of an expanding universe contrary to what many big bang theorists would have us believe. The chance of that galaxy centrality being the case is one out of a trillion, and yet the scriptures predict this is what we would find to be true. When you get the right science based on God's worldview, everything begins to fit together into a logical whole.[19]

MINORAS

We have already seen that on the first day God said, "Let there be light," the word light being "or" in the Hebrew. Now we see in verse 14 on the fourth day God then says, "Let there be lights..." or luminaries as others have translated the Hebrew word "ma'or." The idea of "ma'or" suggests that God was forming light holders on the fourth day that He was transferring His light to, or light sources to replace His own glorious light for the time being.

It needs to be made clear that the sun, moon, planets, and stars were placed in the expanse outside the water canopy and the expanse of the earth's atmosphere below the water canopy.

In verse 6 it says, "Let there be an expanse in the midst of the waters, and let it separate the waters from the waters." This expanse is also referred to in verse 20 where God says, "...let the birds fly above the earth in the open expanse of the heavens." In verses 14-17 God refers to another expanse that is not between the lower and upper waters of day 2. This expanse is above the expanse that the birds fly in. This distinction of different expanses makes it clear that the Bible understood from the beginning that there was a separation between the earth's atmosphere and space where the sun, moon, planets, and stars dwell. This separation of expanses is not obvious to someone looking from earth into heaven, but God's revelation has made it clear. Apparently the word for expanse and the word for heaven in the Hebrew are used interchangeably with expanse being the common term and heaven being the formal term describing any stretched out spheres that are part of God's creative work. Dr. Henry Morris explains this distinction in his book, *The Genesis Record,* on page 67.

These givers of light had a much greater purpose than just to provide light on the surface of the earth. God tells us in verse 14 that the lights were to, "separate the day from the night, and let them be for signs, and for seasons, and for days and years." We have already seen that our solar system, in relationship to the rest of the stars and galaxies, is one big clock. It is a well known fact that all of the ancient civilizations followed the stars in their constellations and developed elaborate meanings to apply to their organized structures (signs). They followed the stars in hopes of getting insight into coming events or meaning from events that happened during the time a constellation was in the sky. The three wise men who came from the East made the journey because of a sign they saw in the stars.

Dr. Henry Morris in his book, *Many Infallible Proofs*, on pages 337-342 gives an extensive discussion of how the original meanings of the Zodiac signs were corrupted after the flood. There is a great deal of evidence that originally the stars told the story of God's redemptive plan to send His Messiah through a virgin (Virgo). This Messiah would finally become the King (Leo).[20] Dr. D. James Kennedy in his book, *The Real Meaning of the Zodiac*, gives an even more extensive description of this gospel in the stars theory. He makes the point that every ancient civilization from Rome to Babel used the same designations for the constellations and that no one knows their real origin. Dr. Kennedy shows those origins to come from God through Adam.[21] The wise men of the East who were star gazers certainly had some information available to them that suggested a king was going to be born in Judah when they came to Jerusalem asking about that king at Herod's palace. They had seen His sign in the east.

The idea of seasons also had a much broader meaning than just knowing when to plant and when to harvest. Before the flood, we have already noted, the growing season and harvest season took place year round in the fairly even temperatures caused by the vapor canopy. Now these seasons would refer to knowing the right times to give sacrifices and have religious celebrations. The Sabbath day of rest was established from the beginning of creation. Cain and Abel brought sacrifices to God at a certain time. Later in the law, God established special feast days that were to be kept during specified seasons of the year. As the people on earth watched the stars, they could look forward with anticipation to their next coming season of worship and celebration. One thing is very clear. God intended for man to be able to read the stars in relation to Him as Creator of all things, including the sun, moon, and stars.

By what we read in Genesis 5 about the life spans of the early descendants of Adam, we learn that Adam and his descendants could read the stars from the beginning, and what they saw during their time is much the same as what we can see in our day by the naked eye. One thing we can know for sure is that the precise and predictable arrangement of the stars cannot be the result of a random chaotic explosion. There is no chaos in the heavens.

ALIENS

One thing made very clear by what we read about how God created the earth first and then the sun, moon, and stars is life was only created on earth. There are no aliens out there on other planets orbiting other stars in their own solar systems similar to ours. The earth is God's focus of creative activity, especially when it comes to life. In this day and time, we make such a big deal out of outer space and going to the stars with Star Trek or Star Wars visions dancing in our heads. The creation account makes little to do over the stars. In verse 16 they are mentioned almost as an aside when it says, "He made the stars also." The Bible does not even mention other planets besides earth. The making of the stars was no big deal to God. They are made up of mainly hydrogen and helium that are big balls of fire. The earth is far more complex in its make-up and position in space.

Granted, the stars are arranged in obvious designed patterns and in perfect relationship to one another in an order that becomes more exquisite the more we learn about them. But, none of that compares to the minute details that make up the complexity of the earth (including life forms) with all of its intricately interwoven systems that depend on each other for their existence.

Where did the concept of aliens, as we know them today, come from in the first place? The evolution worldview's cosmology tells us we humans are insignificant beings on an insignificant planet in an insignificant solar system with a sub-par star for a sun in a minor galaxy amongst billions of other galaxies. All that we see and are is nothing but the result of random events that happened to come together by chance out of a chaotic explosion. On the basis of all the other stars, planets, and galaxies that are out there, who is to say that similar events could not have taken place in other parts of the universe that resulted in other life forms. It happened here, why not on other planets as well? Aliens are a logical conclusion to evolution theory.

The problem is, no messages have ever been received by all the billions of dollars worth of equipment that has been developed for the expressed purpose of listening for intelligent signs in space. The government spent billions of tax dollars on SETI, Search for Extraterrestrial Intelligence, and finally wised up and made these alien chasers go after private money from donors like Stephen Spielberg and George Lucas to name a few. Much of space exploration is more about finding life on other planets than it is about increasing our knowledge. Why is this so?

To discover life on other planets would be the strongest evidence for evolution ever discovered, at least in the eyes of the evolutionists. It has been amazing to watch the scientists sending robots to Mars. The planet is desolate and yet they look for any shred of evidence that water once existed there. Water means the possibility of life and finding life on Mars means evolution has to be true in the minds of evolution scientists.

Let me make it clear though, that finding water anywhere in space does not mean life exists in that water or because of that water. We know ice exists in Saturn's rings, on one of Jupiter's twelve moons, and comets are made of ice. So, there is water in other parts of space but that water gives no evidence of life. It takes more than water to produce life. When God made the sun as a light holder, the moon and planets as reflectors, and the stars as light holders, they were to provide light for earth, not sources for life in other solar systems.

A YOUNG UNIVERSE

There are several evidences that the universe as well as the earth are young. When we look at the pictures from the Hubble telescope in space, we see vast clouds of dust surrounding stars in galaxies. If the universe is billions of years old, that dust should have been swept up by the gravitation of the stars the dust envelops. The ice rings around Saturn are inner woven like weaving in a basket. Due to the strong gravitational pull of Saturn, those rings should be smooth by now, if Saturn is billions of years old. One of Jupiter's moons, Io, is not much larger than our moon. It is still bubbling with volcanic activity. Because of its size and the voracity of its volcanos, it should have burned out millions of years ago, and yet, it shows no signs of cooling off any time soon.

The moon has been pulling away from the earth at a rate of one and one half to two inches a year. If the earth is only six thousand years old, then it would have moved only several feet. But if the earth is billions of years old, the moon should be of sight by now.

The sun is also shrinking in size at a rate of five feet an hour. Again, six thousand years would not make much difference in its size, but if you add all that shrinkage back over a period of just 100,000 years the sun would be twice as large as it is now. If that were the case, the heat on earth would be so hot that no living organism would be able to survive. Remember, evolution tells us the dinosaurs died out sixty million years ago, not 100,000 years ago. There are also star clusters that are made up of stars moving away from each other at fast rates. Those rates tell us they can only have been separating for thousands of years, not billions of years.

One other thing, the earth is pelted with meteorites at a fairly constant rate each year. The number of meteorites found on the surface of the earth suggests they have been accumulating for only a few thousand years. Even more telling then that is meteorites are seldom if ever found in the rock strata and coal beds of the, so called, geological column. If the flood took place in only a year's time, you would not expect to find meteorites in the flood sediments. If they were accumulating over billions of years, as uniformitarianism would imply, they should be abundantly found in every layer of strata. They are not there.

The above data is only a small part of all the data that supports the Genesis account of creation that took place a few thousand years ago in a six day time period that was then totally changed by a world wide catastrophic flood.

The evidence strongly supports the Biblical account that God created the sun, moon, planets, and stars in one day. He spoke according to His will that they be formed and placed in patterns that can be used to tell exactly where the earth is, has been, and will be in space and time.

CHAPTER 7
WHICH LIFE?

At the end of the fourth day, God had finished all His preparations that anticipated the next level of His creative activity. He established the time, space, matter continuum and then energized particles into atoms to form elements that formed into compounds, such as water, the atmosphere of gases, the protective canopy, land, and vegetation. In the space surrounding the earth and its canopy, He made the sun, moon, planets, and stars to provide needed light for life and measurement of time. In the process, He pronounced each new making of the elements as being good or a perfect expression of His person as the Creator.

This was the first level of *bara'*. God's creative activity established in the phrase of verse one, "In the beginning God created..." On the fifth day, we see that the word *bara'* translated "created," used once again in verse 21 in the phrase, "And God created the great sea monsters." God has now entered into His second level of creative activity; that of creating life in a vast variety of physical body forms. Unlike plants, these bodies will have the blood of life (Leviticus 17:11&14) or *nephesh* that will allow them to have various levels of awareness of their existence as living organisms.

The Genesis record makes a clear distinction between plants and animals. Plants are not living organisms. They do not possess the *nephesh* or animated life that animals and man possess. When a plant is harvested for food, it does not die because it never was alive in the way scripture defines living things. Plants do not possess *nephesh*.

SEA MONSTERS

God has brought into existence the hydrosphere (water) on the first day, the atmosphere (sky) on the second day, and the lithosphere (land) on the third day. Now His plan calls for a complete filling up of each of those three earthly spheres with life. This is a wonderful testimony to the fact that God is the living God Who is all about giving life. It is no wonder that this great God of life cries out to Israel as they prepare to enter the promised land in Deuteronomy 30:19, "So choose life in order that you might live, you and your descendants." God is all about life. He gets no pleasure out of seeing things die. He sent His Son so that those who believe in Him might live.

We find a pattern here. God formed the water, sky, and land in that order. Now He will create living animals to fill each of those spheres in the same order. He creates animals to live in the water, animals to live in the skies, and then animals to live on the land. As we have studied these many different life forms, it is amazing to see the wonderful ingenuity of the Creator engineering each kind of animal to be specifically adapted to its sphere of habitation. Not only are they fully adapted to their habitat, but they are vast in variety of shapes, sizes, and colors. The animal kingdom truly does express the glory of God's creative genius in all its wonder.

God begins His creating of living creatures with those who will inhabit the seas of the earth. At least evolutionists get one thing right. The first animals created were those that live in the seas. But it seems they have gotten it all backwards. Evolution says animals evolved from small simple one celled creatures to larger and more complex creatures.

Genesis 1:21 tells us that God began with, "...the great sea monsters, and every living creature that moves, with which the waters swarmed after their kind..."

This verse tells us God started with the largest of animals and then moved to smaller animals. With the study of viruses, single cells, and animals on the lowest level of the geological column, scientist have come to learn that there is no such thing as simple organisms. Viruses and single cells have as much chemical activity going on within them as a city the size of New York. The trilobite had an eye more complex than the human eye. The fossil record makes it very clear that they were all fully formed from the beginning.

Charles Darwin wrote of his version of the Tree of Life in his presentation of evolution. It starts with one single celled animal at the bottom of the tree trunk and then branches out showing smaller sea animals that branch out into amphibians, then into reptiles, then into mammals that develop into man at the top of the tree. There is a problem here. The fossil record does not support Darwin's Tree of Life concept at all. Instead, the fossil record shows an explosion of all kinds of life forms from the very beginning. They call this the Cambrian Explosion. The Cambrian level of the geological column is the lowest level where life forms are first found. In the Cambrian, contrary to evolution theory, they find molluscs, soft bodied jelly fish, exoskeleton life forms, and endoskeleton life forms. Every phyla in the designation of relationships of animal life forms are represented in the Cambrian Explosion. All these animals were complex life forms that show no previous ancestor. All of a sudden they are just there. To make matters worse for evolutionists, even human relics have been found in Cambrian strata, like a sandal and foot prints in Antilope Springs, Utah , and iron bands in Lochmaree, Scotland.

We will discuss the Cambrian Explosion more in depth when we get to evidences of the flood, but suffice it to say at this point in our study that the fossil record supports the Genesis record that called all created life forms into existence on the fifth and sixth days of creation.

The creation model predicts the Cambrian Explosion. The evolution model must adjust its assumptions to explain it.

Some evolutionists try to say that animals before the Cambrian age were too small or soft bodied and were not preserved. They existed but were not fossilized. In contrast to that argument, studies have discovered microscopic fossil bacteria in pre-cambrian layers. There have also been soft bodied fossils found. These fossil remains tell us that, if pre-cambrian animals existed they should be recorded in the pre-cambrian sediment regardless of their physical make-up. They are not there. There is no getting around the Cambrian Explosion for evolutionists. It is a beam in their eye that cannot be removed.

The *King James Version* calls the great sea monsters of Genesis 1:21, "whales," but the Hebrew word "*tannin*" is translated "dragon" and "serpent" in the rest of scripture. Now whales are impressively large, but they are not what we would consider to be dragons. This is especially true since, in the last few hundred years, so many large and much more ferocious sea creatures have been discovered in the fossil record. The dragons described in ancient literature are not anything like whales. One such animal is described in Job 41 that fits the description of a dragon.

The sea monster is called "leviathan." Many Bible commentaries say this animal was probably a crocodile, but the description of leviathan given in chapter 41 of Job tells us this animal was far different from crocodiles that still exist today.

The book of Job is a very interesting book. Studies of its content such as names, places, cultural practices, and such tell us the book is contemporary with the time of the patriarchs Abraham, Isaac, and Jacob. That being true, then Job lived just a few hundred years after the flood.

That would mean that many of the animals extinct today would have still been in existence in Job's day. The only reason why a commentator of the book of Job would call the leviathan a crocodile is that a crocodile is the closest thing living today to the leviathan Job describes. It is obvious that God, in His rebuke of Job, uses the leviathan as an animal that expresses God's great creative power. He chagrins Job by reminding him of His authority over all of His creation, including the leviathan, an animal so powerful that man could not control it.

It is also understood that God describes an animal Job was aware of and had possibly seen. This animal was so large and ferocious that it invoked fear in Job's heart just to think about it. We are also told it had tightly knit scales, and breathed fire and smoke, things not characteristic of crocodiles. The leviathan better fits the description of dragons we find in antiquity then present day crocodiles. It is also very likely that the *tannin* created in Genesis 1:21 is referring to the leviathan of Job 41.

There are two points to be made here. First, there is a principle in the study of mythological stories. There are implications that myths are distorted stories of what actually happened in the past but were embellished over time as they were passed on from generation to generation. Job's leviathan tells us dragon type animals did exist in the past. Dinosaur fossils tell us the same thing. Second, many such animals survived for a period of time after the flood before dying out or being exterminated by man or changing weather patterns after the flood.

That would imply to us that animals we find in the fossil deposits were on the ark. We will discuss this further when we get to our study of Noah's ark. My point is that just because we do not see animals today with certain abilities like breathing fire does not mean they did not exist in the past. It only means we have not seen them at the present time.

Think about it like this. It would be hard to believe animals could produce their own electricity if we had not seen electric eels. What about fireflies that produce light by non-heating bioluminescent without electricity or bombardier beetles that produce a spray out of their tail that is as hot as boiling water? You would have a hard time believing they existed if you had not seen one.

Cows eat grass. When they belch, they belch methane gas. Methane gas is highly flammable. That is why scientists are looking for ways to produce it on a large scale to use it to replace fossil fuels for gasoline production. All a cow would need is a mechanism to spark the methane gas it belches to make it into a fire breathing Guernsey. It is not to hard to conceive that an animal that ate large amounts of foliage could breath fire if it had a mechanism created in its anatomy that would allow it to ignite the methane gas it belched at will. With these things in mind, we are allowed to suggest that the great sea monster, "*tannin*," was just that, a fire breathing dragon. Later we will see how this serpent dragon became the symbol of Satan as we read in Revelation 12.

A final word about sea creatures. The so-called, extinct wollemi pine tree mentioned in the last chapter, brings up another, so called extinct fish. The coelacanth was thought to be extinct for the last 70 million years.

It was used as an "index fossil" to help date rocks. That means that a rock formation that had a coelacanth fossil in it was 70 million years old. The problem is coelacanths have been caught alive since the 1930's. This fish was once thought to be a transitional fish evolving toward developing legs for land use because of its unique arrangement of fins. The discovery of living fossils like the wollemi pine and the coelacanth have brought into question the validity of the Geological Column and its use as a tool for dating rocks and fossils.

The Geological Column was established over 150 years ago when far less was known about the fossil record than now. In reality, it is a product of much speculation with no real scientific means of verifying it. It is a figment of Charles Lyle's imagination, the man who popularized the column.

BIRDS

After filling the waters with all manner of living creatures, God then created the fowl of the air. These special animals would have the ability to fly in the atmosphere. They would require a very different kind of physical make up than the creatures in the waters. Things like a lighter hollow bone structure than that of other animals; eyes that have the ability to see for long distances from the air to the ground; beaks that cut through the air while in flight; feet that would allow them to stand on tree branches and also fold into their bodies during flight; and feathers that would give support in flight but also make it possible to fold their wings when not in flight. Of course, there were mammalian bats and reptilian pterodactyls that also flew with wings that had skin like membranes covering their bone structures rather than feathers.

There are snakes and squirrels that have the ability to maneuver from tree to tree through the air but their flight is more of a gliding ability rather than true flight. Many forms of insects also have the ability of flight. All of these kinds of flying animals would have been created on the sixth day mainly because they do not fit in the bird category referred to in the text.

It seems that God created some animals that cross over lines of distinction in order to produce variety and show His versatility. We find birds that have feathers that do not fly and squirrels that do not have feathers and yet they can glide like a bird. We have already mentioned bats and pterodactyls.

This frustrates those who try to support their evolutionist theory with clean cut schemes showing species naturally changing into different species but maintaining certain similarities. Instead, we have species that are each unique to their own kind with no intermediate transitional relatives.

Birds are a real problem for evolutionists. How did they develop their ability to fly? Flying is very complex behavior. There needs to be enough wing structure to support lift; there must be adequate control surfaces; the body must have a light bone structure; there needs to be a special circulation to provide ample oxygen to support a very high metabolism; they must have instincts to help them navigate air currents and varied wind velocities; and so on. How did feathers develop?

Feathers are so distinctively different from hair, fur, or scales. How did feathers evolve? Feathers are feathers in the fossil record. There are no animal forms in the fossil record that show any evolutionistic development of feathers from hair, fur, or scales.

The popular theory at this time is that birds evolved from two legged dinosaurs. They claim these dinosaurs had hair that later evolved into feathers. There is no solid evidence of this.

There are some dinosaurs that are lizard hipped while others are bird-hipped. The problem is that many two legged bird hipped dinosaurs are older, according to evolution dating methods, than bird-hipped four-legged dinosaurs.

You would think that four-legged bird-hipped dinosaurs would have evolved into two legged bird hipped dinosaurs and then into birds. In many cases the opposite is true. One example is the more recent triceratops.

This huge four-legged animal was bird-hipped, not lizard-hipped, and yet evolved very late. It is a better example of evolution in reverse rather than in progress. Their fossil record interpretation is backwards. Maybe birds were first two-legged dinosaurs with no hair then four-legged dinosaurs with hair that became two-legged dinosaurs with hair that finally developed feathers and then learned to fly after their bone structure totally changed. Just as this total speculation is absurd, the suggestion that birds evolved from dinosaurs, with no real hard evidence to support the idea, is just as absurd.

The one fossil evolutionists use as their example of a missing link between dinosaurs and birds is the archaeopteryx. This extinct bird had teeth in its beak, claws in the middle joint of its wings, and a tail with feathers on it. Many of those who have studied this rare fossil have come to the conclusion that it is nothing but a bird. Every characteristic of the bird, like teeth in its beak or clawed wings, are found in other species of birds. It is an extinct fossil but not a missing link between dinosaurs and birds.

The archaeopteryx was also found in sediment rock that was determined to be older than some of the dinosaurs it was supposed to have evolved from. This is another example of evolution moving in the wrong direction. A much better explanation is all these animals died at the same time in the flood and are now found in different levels of strata that was laid down in the same catastrophic event. They all lived before the flood and then became extinct after the flood.

Here is the real problem birds give evolutionists. Genesis 1 tells us birds were created on the day before dinosaurs were created. According to creation, birds predate land animals by one day. It also tells us that the birds, like all the plants and animals were created according to their own kind.

They are all DNA specific, and thus genetically bound to reproduce their own kind. It is amazing to me that this idea of animals changing from one kind to another still persists after all the years of studies concerning DNA. These studies, when presented honestly, continually reaffirm the fact that any mutative change in the DNA molecular structure is a loss of information or distortion of information. This loss of information causes defects in the animal not progressive change from one kind to another. The only real reason for this speculation to persist is the fact that those who refuse to accept the Creator have no other alternative than to keep holding out hope that something will come along to support their faith assumptions of evolution. Dr. George Wald, winner of the 1967 Nobel Peace Prize in Science once admitted:

> "When it comes to the origins of life on earth, there are only two possibilities; creation or spontaneous generation (evolution). There is no third way. Spontaneous generation was disproved 100 years ago (by Louis Pasteur), but that leads us only to one other conclusion; that of supernatural creation. We cannot accept that on philosophical grounds (personal preference); therefore, we choose to believe (by faith) the impossible; that life arose spontaneously by chance." (The parentheses are my commentary)[22]

I appreciate Dr. Wald's candid honesty. If other natural evolutionists were as honest, they would all have to admit, like George Wald, that their commitment to evolution theory is more about rejection of God than being committed to true science.

We have already mentioned pterodactyls, those huge flying reptiles. One such flying dinosaur was found in Texas in 1972.

They called it Quetzalcoatlus and it had a wing span of 48 feet, which is wider than a F4 phantom jet. That is one big flying machine. Such a large flying animal would definitely require a heavier atmospheric pressure to help it become air born. I saw a documentary on television not long ago that reported that some eagle or hawk like feathered birds have been found in the fossil record that were large enough to carry away a medium sized human being. I was not surprised by the report because the creation model predicts these are the larger kinds of animals that lived before the flood that you should find buried in the strata left by that same flood.

There are several animals living today that are found to be much larger in the fossil record besides these birds- animals, like beavers, sloth, bison, bears, and even King Kong sized gorillas (well maybe not quite that big.) These huge apes are called gigantopithicus. Some speculate they still exist but are misrepresented such as the well known and often sighted Big Foot.

Large size is true of insects as well. Dragon flies with eighteen inch wing spans and cockroaches the size of small mice have been found in the fossil record. In the text concerning the fifth day, God calls for the waters to be filled with swarms of creatures and then calls for the atmosphere to be filled with swarms of birds. All of the swarms of animals were then mandated to be fruitful and multiple and fill the earth according to their own kind. This suggests that there was an abundance of food and an abundance of space to fill. God wanted to fill these spheres of existence with life to express the fact that He is the God of life. It was all pronounced good as another level of His creative plan perfectly expressing the kind of God He is; a God of life, of abundant provision, of great variety, and of wondrous beauty.

LAND ANIMALS

We now come to the first part of the sixth day. The next step in God's sequenced plan of creation was to fill up the land with living creatures all according to their own kind. To accomplish this objective, God forms these living animals on three levels, beasts, cattle, and creeping things on the ground. Genesis 1:24 says, "Let the earth bring forth living creatures after their own kind: cattle and creeping things and beasts of the earth after their own kind." In the next verse, we are told that the sequence of God's work started with the beasts, then the cattle, and then the creeping on the ground animals. This is consistent with the approach He took when creating on the fifth day. He started with the larger animals and then moved to forming the smaller animals. This approach to designating the land animals has no correlation with the present way of designating animals by biologists and zoologists who use designations such as amphibians, reptiles, mammals, and insects.

Rather, they are described more in terms of large animals that can graze in the upper levels of the vegetation, medium sized animals that would graze in the mid-levels of vegetation, and then the creeping things that would find their food in the lower levels of vegetation. All these sizes of animals would be needed to keep the ecosystem in tack. Their grazing would allow the vegetation to get all the sun light it needed. This continual grazing would also keep the vegetation eaten back so it did not completely over take the earth. The year round growing season and benevolent weather patterns caused by the vapor canopy would cause massive vegetation growth. Large amounts of animals of all sizes would be needed to keep the vegetation eaten back. It also seems that the use of the word "cattle" would suggest that some of these animals would become those most appropriate for domestication. Other mid-range sized animals probably would not be used for domestication like some deer, antelope, and others as it is today.

The term "creeping things" would mean all animals that live close to the ground including small mammals, reptiles, amphibians, and insects. The insects are a world unto themselves. Their varieties are vast. One thing you do not learn about insects from evolutionists is they are much the same in the fossil record and amber as they are today. For example bees we find in amber that is supposed to be millions of years old are very much the same as the ones that exist today. Pierre Grasse' wrote in his 1977 book, *Evolution in Living Organisms,* page 30, "We are in the dark concerning the origin of insects." That means there are no transitional forms of insects. Flying insects are an even greater problem because there is no similarities between their wing structures or the way they fly. They all seemed to have developed out of individual patterns rather than a common ancestor.

The common fly, beetles, and butterflies are all totally different in wing structure and process of flying. Evolution has no explanation for all the diversity amongst insects. Creation, on the other hand, understands that each was created according to its own kind from the beginning just as the fossil record affirms. The group designated as beasts is intriguing. Obviously, these were the larger animals that would live in the wild and not as subject to man's control as those designated as cattle. It is intriguing because the largest animals we find in the fossil record of animal life are the dinosaurs. Those "terrible lizards" as the name dinosaur, means in the Greek.

The evolution worldview that believes dinosaurs died out millions of years before man's introduction into the evolution scheme has, for sometime, been a source of resistance to accepting the creation account literally. The continual brainwashing by secular science books, documentaries, movies, museums of natural history, journal reports, and so on have almost totally indoctrinated our whole society into automatically thinking of millions of years when one sees a picture or replica of a dinosaur.

This being true, it takes a person a while to let the fact sink in that, if God created all the animals on the fifth and sixth day, that creative activity would have to include all the dinosaurs, not to mention all the other extinct animals like mammoths, saber tooth tigers, and giant rhinoceros found in the fossil record. The creation worldview makes it clear that all the animals found as fossils in the layers of geological strata had to have been created during the creation week. We have no accounts of God creating before Genesis 1 and the account itself tells us God did no more creating after day six.

This means that the creation model predicts that man and all animals lived together at the same time. This being true, then evidence should be found to show that was the case. The creation model also predicts that all those land animals would had to have been on the ark and survived for a time after the flood before dying out. That is, at least all animals in their original kinds before diversification began to set in. We will discuss this further when we get to the number of animals on the ark. We will see that these predictions are supported by what we can observe.

DINOSAURS AND MAN

The Bible and the fossil record make it clear that humans and all animals lived together at the same time, including the dinosaurs. We have already seen that Job 41 describes a dragon like animal called "leviathan," that would fall into the category of dinosaur if its remains were discovered today. As a matter of fact, there are some dinosaurs that lived in water and on land that had a cavity in their skulls connected to their nostrils not found in other dinosaurs that may have contained an apparatus to possibly ignite methane gas and make the animal fire breathing.[23]

But Job also introduces us to another animal that fits the description of the long necked, large tailed brachiosaurus pictured in the movie Jurassic Park or the equally as large apatosaurus that grew to dimensions similar in size to the largest sea creature known to man, the blue whale. These animals grew to 100 feet long and weighed up to 100 tons, such as the one uncovered in western Colorado referred to as Ultrasaurus. Other similar finds have been named Argentinasaurus, Supersaurus, and Sizmasaurus.

There has been some debate whether some of these large long necked and long tailed dinosaur fossils were brachiosaurus or apatosaurus but they definitely are the largest land animal fossils found to date. Some of these sauropods would have survived for a while after the flood before becoming extent in the much more volatile post flood environment. Of course juvenile pairs would have gone on the ark that were not fully grown and would not have required nearly as much space or food as an adult.

Job 40:15-24 introduces us to the land animal called "behemoth." Modern translations try to tell us that this was either a hippopotamus or an elephant, but there is a problem with those suggested translations of behemoth. Neither one of those animals fit the description of behemoth given in the text. The real give away is verse 17 that says, "he binds his tail like a cedar." Now a cedar is a large tree not a short bush. The hippopotamus and the elephant both have small tails, but brachiosaurus had a tail as big as a cedar tree. Another interesting point made about behemoth in Job's account tells us this huge and extremely strong animal was unmoved by the river Jordan when it ragged during the flooding season. It says that, "He is confident, though the Jordan rushes to his mouth."

This suggests the behemoth could stand on the bottom of the river and, even when it was at its deepest time of the year, the river did not rise enough to cover the animal's mouth.

In the case of the brachiosaurus with its nostrils on the top of its head, this animal could keep his mouth under water and still be able to breath out of the top of his head. It would take an animal with a very large body and long neck to accomplish these feats described in Job. My point is that God describes to Job an animal that Job had seen living at the same time Job did.

It was one of God's examples of His great creative power to remind Job of his own smallness in light of that creative power. In our chapter on the post flood era we will look at other evidences that suggest dinosaurs and man lived together at the same time and some dinosaurs lived several years after the flood in different parts of the world. Suffice it to say at this time that human tracks have been found along side dinosaur tracks as well as in dinosaur tracks, suggesting those tracks were made at the same time when the ground was still soft. These combined kind of tracks have been found in Australia, Russia, and Glen Rose, Texas on the Paluxey River. Actually, evidence of human existence has been found in every level of the geological column including the Cambrian. One such example of that evidence is a fossil foot print made by a human sandal with laces around its edge that has a trilobite under the print showing the trilobite was stepped on in the mud by a human. The trilobite belongs to the Cambrian era and yet this fossil tells us man does too. This kind of evidence is so important because it helps establish that all the animals in the fossil record lived with man and thus were all created at the same time during the creation week as Genesis 1 tells us. Evolution theory cannot stand if man and all the animals in the fossil record lived together at the same time.

That would make it clear that there was no such process of animals evolving from simple to complex life forms over millions of years. They all came into existence at the same time according to their own kind.

It is interesting to note that God made this point to Job when He said to Job in Job 40:15, "Behold now Behemoth, which I made as well as you..." or "when I made you..." Understand that God was speaking in the context of the sixth day of creation in Genesis 1, referring to the fact that man and behemoth were made on the same day.

A bit of evidence that comes from the fossil record has been reported as a result of the RATE reports from ICR in San Diego, California. It has to do with dating methods used to date fossils. Carbon 14 has a half life of 5700 years. That means that it deteriorates at a rate that half as much Carbon 14 can be found in a fossil bone every 5700 years until it finally disappears all together. That being the case, no fossil older than 250,000 years should have any Carbon 14 left in it. The fact of the matter is that all fossils, no matter how deep they are found in the geological strata have traces of Carbon 14 in them. The vapor canopy also suggests to us that there would not have been as much Carbon 14 in the atmosphere before the flood then there is now, suggesting there was even less Carbon 14 in the animals before they died. That means none of the fossils can be older than 250,000 years but are really much younger than that (a few thousand years), and all died at approximately the same time.[24] Creation predicts this would be the case. Evolution has to explain why dinosaurs that supposedly died 60 million years ago still have Carbon 14 left in their remains when it should have all been gone long, long ago. It truly implies that evolutionists need to finally admit that their model is wrong.

There are two other discoveries about dinosaurs that need to be included here. Dr. Mary H. Schweitzer of North Carolina State University found the remains of soft tissue in a supposedly 68 -70 million year old Tyrannosaurus Rex fossilized leg bone buried in eastern Montana.

The soft tissue was small microscopic cells found in small vessels in the leg bone.[25] Dr. Schweitzer stated in response to her find, "I am quite aware that according to conventional wisdom and models of fossilization, these structures are not supposed to be there, but they are. I was pretty shocked."[26]

The reason why Dr. Schweitzer was shocked is soft tissues do not survive millions of years of burial. Her find suggests the T. Rex died thousands of years ago, not millions. The other discovery was made by a paleontologist in India. Grasses according to evolution did not evolve until after the large dinosaurs died out. The problem is this paleontologist found the remains of grasses in dinosaur dung.[27] In other words, grass and dinosaurs lived together at the same time just as man and dinosaurs did, which is exactly what God's word has told us from the beginning.

KIND

Like the plants, we see that all of the animals in the water, atmosphere, and on land were all created according to their own kind. "Their own kind" tells us that each animal has its own specific seed for reproduction and each kind can only reproduce its own kind and no other. If that is true, then creationists would predict that all living forms are genetically specific and their DNA would be specialized without the capacity to produce any other life form, except its own.

On the other hand, evolutionists would have to predict that genes and their DNA would have the capacity to transform themselves into other arrangements that would lead to changing one kind of animal into a whole new kind of animal by random chance events. This process is called "macroevolution." Neo-Darwinism implies that evolution takes place within the realm of genes and DNA.

Original Darwinism only concentrated on those physical similarities between species that suggested possible links between species using homology (such as similar bone structure or similar structures in fetal stages of various species.)

It also pointed to changes within species that were caused by adaptations to a changing environment. The most noted example being the changes in the beaks of Darwin's finches on the Galapagos Islands. The finches with larger beaks survived while those with smaller beaks diminished during drought when food became harder to come by requiring larger beaks to break open harder seed shells. When the rains came again the finch population would then reduce in beak size just as before the drought came. They reversed in their microevolution process of adaptation showing no sign of macroevolution taking place.

Neo-Darwinism took over when it became clear that the original supports for Darwin evolution only explained microevolution but did not give evidence of macroevolution. Microevolution observes that there can be variations within a species like color and size but a dog is still a dog and a cat is still a cat. Neo-Darwinism began then, to focus on the genetic make-up of cells with the hope of finding the mechanism for macroevolution there. That is turning out to be a dead end street as well.

Though there are many similarities between the make up of DNA between animals of all kinds and man, even when there is only a difference of 2% to 3% in the number of match-ups between the DNA and RNA molecules, such as in apes and man, that difference still involves millions of totally different amino acid arrangements. The more studies they do on DNA match-ups of man and apes (e.g. heterochromatin) the larger the percentages of differences are becoming, such as 10% to 20%.

The only real mechanism for change is mutation, and there has been precious few mutations shown to bring a positive change for the animal that has the mutation.

And just as important to remember is the fact that, a mutation causes the loss of information not the addition of new information. Loss of information is de-evolution not evolution. Most scientific studies on DNA done to prevent diseases tries to correct miss information in the DNA not to add new information. The need is not for more information in the DNA, the need is for the right information that has already been established in healthy DNA that is not mutated. The reality is that DNA in each living creature has its own original form (design). Any additions or subtractions to that original form by mutations always lead to negative results.

Add to that the fact that DNA has its own correcting mechanism, which works to correct bad information, then you have an impossible hurdle for neo-Darwinism to jump over. The point is that neo-Darwinism's hope of finding the mechanism for macroevolution in the genes of cells is another dead end. Lynn Margulis, famous for her work in symbiogenesis, was a featured speaker at the World Summit on Evolution Galapagos 2005. To the surprise of many, Dr. Margulis questioned the validity of neo-Darwinism on the basis of her research. During her speech she is quoted as saying, "It was like confessing a murder when I discovered I was not a neo-Darwinist." Why could she no longer consider herself a neo-Darwinist? Her study of genetics did not support macroevolution theory. The cells of animals are genetically specific according to their own kind. Again, her own research did not support neo-Darwin theory.

One of Darwinism's big, so called, evidences for animals evolving from one species into another species is homology.

They claim similar structures, such as bones and gullets, plus position of structure, equals common ancestry. Once again, genetics lets them down.

If evolution is supported by homology, then similar genes should produce similar structures and positions in animals. Studies of genes common to animals of different species have shown that those common genes do not always produce the same structures during development. Often, they develop very different structures showing no link between genetic make up and homologous structures. On the other hand, some very similar physical structures in animals form from very different genes.

The problem goes even one step further. Many similar structures, such as the gullet in vertebrates, do not have the same development pathway during embryonic growth. If they all come from a common ancestor, then their embryonic development should be the same. The fact is that they are often not the same. For example, sharks, lampreys, and frogs are all vertebrates with similar gullets. The shark's gullet forms from the top of the embryo. The lamprey gullet forms from the bottom of the embryo and the frog gullet forms from the top and bottom of the embryo. They are similar in form but each of the developmental pathways of their gullets are totally different.

Another example is salamanders. Most vertebrate limbs and digits develop in the embryo from tail to head. Salamanders develop in the opposite direction from head to tail. They are amphibians and yet their developmental pathway for digits is totally different from other amphibians such as frogs. So much for homology from common ancestors.[28] Here is an interesting side. Man's closet ancestor is supposed to be apes. We are told that our DNA is very much alike. Dr. B.C. Nelson did a study on blood that brings this assumption into question.

He compared the make up of the blood of several different animals and came to an amazing conclusion. On the basis of similarities of the different blood he studied, he concluded that man is most closely related to pigs, not apes.[29] This is one reason why pig heart valves are used to replace human heart valves rather than ape heart valves. The human body does not as readily reject the pig valve in most cases.

God does not miss a thing. He looked ahead before He created and saw that man, in his rebellion, would try to say humans evolved from lower animals. So, God made them each in a way (according to their own kind), that when man finally learned about genes, God would be shown to be true and man once again mistaken. It will not be the first time. The Sadducees in Jesus time did not believe in a resurrection and tried to make Jesus look foolish for teaching one.

Jesus showed them in Luke 20:34-38 that scripture teaches God is the God of the living not the dead. He then later rose from the dead Himself and established the fact once for all. Man was mistaken then and will be mistaken again. Why? When you have the wrong worldview, you will be wrong about a lot of truth that your wrong worldview will not allow you to accept.

CHAPTER 8
WHICH MAN?

The Genesis account of the creation of man is very different from evolution based interpretations of the fossil record. As we have already seen, Genesis tells us that man is a special creation; a higher *bara'* than the rest of God's creative work. In the Genesis record, man is not an animal, he is a separate creation from the animals. The animals have life but man has life that is infused with the very breath of God (Genesis 2:7.) Is the creation account true or is the evolution theory true? One is true or the other is true but they both can not be true because they are opposite concepts.

Natural evolution says man is just another development in a long series of chance mixtures of positive mutations coupled with abilities of previous life forms to adapt to the changing environment. All this led to ever increasing complexity until finally modern man has made his way to the top of the food chain, at least for now. Theistic evolutionists agree literally with that description, but say that God or a god guided the process. The problem they run into is the fact that evolution theory was developed as a naturalistic explanation of existence that does not need a god. Theistic evolutionists accept a science that does not accept them. Asa Gray, a professor of Botany at Harvard, was a major contributor early on in the evolution movement in America. He was the first to suggest the concept of theistic evolution to Darwin.

Darwin rejected the concept flatly in a letter he wrote to his evolution mentor, Charles Lyell, in 1861. Darwin wrote, "The view that variation has been providentially arranged seems to me to make Natural Selection entirely superfluous, and indeed takes the whole case of the appearance of new species out of the range of science."[30] This statement makes it clear that as far as Darwin was concerned, God and evolution did not mix.

One of the major problems that confronts naturalistic evolutionists is brought up in the discussion of the creation of man but is true in all of the animal kingdom as well. Verse 27 says that God created them, "male and female." Here is the problem. Naturalism teaches that life began with simple single cells. In the cell level of life, we find that the cells divide without a sexual reproduction process. They are asexual. The two questions evolutionists have to answer are, "When did the process of bi-sexual reproduction begin and why would it have been more efficient than asexual reproduction?"

The fact of the matter is there is no reasonable answer to these questions. Strictly speaking, first, asexual reproduction would be more efficient in a purely survival of the fittest environment; and second, bi-sexual reproduction is so complex it would take thousands of intermediate organism to go through the developing process before there was a male and a female capable of procreation together. What went on in the mean time and where are the fossils of all those animals in the process of developing the bi-sexual ability?

George C. Williams made this observation about sex and evolution, "The problem has been examined by some of the most distinguished of evolutionary theorists, but they either failed to find any reproductive advantage in sexual reproduction, or have merely showed the formal possibility of weak advantages that would probably not be adequate to balance even modest recombinational load.

Nothing remotely approaching an advantage that could balance the cost of meiosis has been suggested. The impossibility of sex being an immediate reproductive adaptation in higher organisms would seem to be as firmly established a conclusion as can be found in current evolutionary thought."[31] Sexuality in naturalistic evolution theory makes no sense. But, if God created us for relationships rooted and grounded in intimate family ties, then sexuality is a must. Progressive creationists, in contrast to theistic evolutionists, say there was a long period of development that was allegorized in Genesis 1, but documented in astrophysics, radiometric dating, and the fossil record.

This progressive development took place over billions of years. The creation of life started recently in the last 530,000,000 years or so, just as the fossil record (from the evolution interpretation of the geological column) suggests. But their take is that God did special creative activities over long periods of time involving millions of years. Animals were being created from simple to more complex kinds, many dying out long before man entered the process. This concept seems wasteful and expresses a low estimation of the gift of life in contrast to the literal creation model where life is held in high esteem.

God, finally, after millions of years of creative trial and error, created Adam and Eve. These humanoids, though, were given the special capacity of spiritual awareness 6,000 to 50,000 years ago. Their figures for dates come from carbon 14 dating methods of religious relics, not the Biblical chronologies.

This is to be expected when we understand that progressive creation uses man's science as its final authority and reinterprets scripture to fit its worldview's "old earth" assumptions. It is an attempt to straddle the proverbial fence.

They strive to be accepted as good scientists in the secular world and good theologians by the faith community. In doing so, they fail to bring truth to either science or theology.

Progressive creationists see the days in the creation week as long ages and do not accept a literal worldwide global flood. The global flood makes the fossil record the result of one worldwide event, rather than the product of millions of years of slow development. That contradicts their "old earth" assumptions. Because each day is an age rather than a literal day, they believe we are still in the Sabbath Day of Rest age that began with the creation of Adam and Eve. Their argument is the seventh day is not said to have an evening and morning like the other six days, suggesting it is not complete yet.

They over look the facts that the number seven is applied to the day just as the other six days and the two "ands" in Genesis 2:2 work as conjunctions connecting the seventh day to the other six days. Exodus 20:11 also speaks of the seventh day in past tense as a part of the creation week that included the seventh day, when God said, "For in six days the Lord made the heavens and the earth,... and rested on the seventh day..."

Progressive creationists also overlook the fact that death was a product of sin and did not become a part of the creation experience until after Adam and Eve sinned and God then pronounced the curse of death on all of creation. Paul says in Romans 5:12, "Therefore, just as through one man sin entered into the world, and death through sin, and so death spread to men..." Progressive creation theory requires that animals were living and dying long before man came into existence. That is a direct contradiction to the Biblical record that requires man to sin before death becomes the curse on all creation brought on by sin. A major Biblical doctrine is man and animals did not die until man sinned.

A creation including death cannot be good because death is not good. Any theory of creation that rejects the literal interpretation of the Genesis record finds itself continually running into contradictions in the rest of the Bible. This is true of progressive creation.

CREATION OF MAN

All the land animals were made in the first part of the sixth day. The vast variety we see in the animals still alive today and those continuing to be found in the flood fossil remains tell us that God would have to have made them immediately as He called them into existence in hosts of living creatures. Each kind of creature would be carrying its own special genetic code that would provide the variables of size, shape, and color within each kind. It would also make it possible for those animals to adapt when their environment would be changed by the flood to come later, making microevolution possible.

We see that God was pleased with all the land animals that He willed into existence as He commanded them to be fruitful and multiply and then pronounced His work in making the land animals "good." But now, God has come to the crowning moment of His creation work. Now He is ready to create the one being He had created all the rest of creation for.

This is the being God will give dominion over all the rest of creation. This is the being that will be the very expression of God Himself in all the universe. This is the being that will be the object of all of God's great love and His invitation for companionship. The one being that will be able to relate to God in a way no other being in creation can relate to Him, including the angles. The time has now come for God to make the declaration in Genesis 1:26, "Let us make man in Our image, according to Our likeness..."

This declaration was made only in reference to man (Adam and Eve) and no other created being. No other being was made in the image of God. Truly, in the beginning, man was to be created a little lower than God (In the image of God) not lower than the angels. The angels were already created to be ministering spirits between God and man (Hebrews 1:14.).

It is amazing to me to see the depths to which this highest achievement of all creation has fallen due much to the misconceptions enhanced by evolution theory. Today, evolution scientists study apes to learn about man's latent animalistic tendencies and then apply their, so called, discoveries to their psychological interpretations of human beings. It is one of the most laughable things I have ever seen in the realm of so called scientific study. In science books and philosophy books, man is often denigrated as another chance result of long sequences of random selection and trial and error survival attempts that finally favored larger brain capacity. This larger brain then, gave the homosapien species power to establish its dominance over other land creatures. Man is just a little more highly evolved animal, in other words. He is nothing special, just one of the millions of accidents of natural evolution. It is no wonder mankind has begun to act more and more like animals.

Low self-esteem is becoming rampant as suicides amongst young people grow out of control. This is no wonder when the false reality of meaningless human existence forces them to the logical conclusion that living life is not worth the trouble. What does it matter if you die and return to oblivion? What does it matter if there is no accountability for your actions or no meaning to your existence? The Genesis account of man's creation tells us that man is not an animal, but a separate creation from the animals.

The physical elements that make up his body are the same as animals but he has been created in such a way that allows man to posses the very essence of God Himself as a part of his own existence. Man is the only being that God individually formed from the dust of the ground and breathed into him the very breath of God's own life (Genesis 2:7). What is true of Adam was also true of Eve. God formed her body from matter taken from Adam's body.

That matter was taken from the dust of the ground and so Eve was also made from the dust of the ground. God would have also breathed into her the same spirit of life that He breathed into Adam. If that was not the case, then the matter taken from Adam would have already received the same life as Adam; but Genesis 1:27 tells us God made man both male and female together. This established an equal level of creation relationship and an equal standing as being made in the image of God. Before the fall, man and woman were co-equal expressions of the image of God.

After the fall, when the environment changed, man became the protector of the woman. This co-equality is restored in Christ just as Paul wrote in Galatians 3:28, "There is neither male nor female; for you are all one in Christ Jesus."

Actually, on the basis of which it is stated in the context of Genesis 1:27, it is the man and the woman created together as a unity that truly expresses the image of God. Verse 27 says, "And God created man in His own image, in the image of God He created him; male and female He created them."

This verse makes it clear that it is the man and the woman together that truly express the image of God. In verse 26, God has already said, "let us make man in Our image..." The scriptures will later reveal to us that God is a tri-unity of Father, Son, and Holy Spirit.

God is one and yet He is three in one, but each of the three are the same. The man and the woman living together in perfect harmony and love toward each other would be a perfect expression of God's unity with Himself. You add the children that would later make up the family, and you then would have a tri-unity. Unfortunately, Adam and Eve sinned before the family could fully express that tri-unity. Man's fallenness aside, the family still has the opportunity to express the oneness of the Godhead when father, mother, and child live in love and harmony with one another.

Man created in the image of God establishes some very important qualities that apply to man. The most important is each individual human being demands respect from every person, regardless of station in life, for each individual is an expression of their Creator. The richest man in the world or the bum on the street are both reflections of the image of God. Being made in the image of God is what the founders of the United States understood gave individuals certain unalienable rights. Those rights were endowed by their Creator. The worldview of evolution provides no such authority for man. Evolution implies that power belongs to the strong and the weak do not deserve to live.

The only authority there is belongs to those who gain power over others. As the saying goes, "Those who rule make the rules." Human beings are not animals. They should never be treated as such. To show disrespect for a human being is to be disrespectful to God Himself.

We will see later that God established capitol punishment in Genesis 9:6 as a right of government to exercise against those who commit murder. The issue there is that a murderer reveals a heart that lacks respect for God who made human beings in His image.

When we no longer fear God with reverence, the taking of human life, without regard for who that human life represents, becomes a natural consequence. The killing of unborn babies along with the old and weak is a sure indicator of when a nation has lost its fear and reverence for God. Cultures who loose sight of God soon find themselves acting more like animals then like human beings made in the image of God. Life becomes cheap.

MAN'S POSITION

In Genesis 1:28, God blesses the man and woman and then gives them three directives. This verse says, "And God blessed them; and God said to them, 'Be fruitful and multiply, and fill the earth, and subdue it; and rule over the fish of the sea and over the birds of the sky, and over every living thing that moves on the earth.'" This blessing gave man the ability to procreate and have many children while living in the prosperity of the abundance of God's provision for them in all the creation. The three directives would be to fill the earth with human life, subdue the earth, and rule or have dominion over all the living creatures that God had created.

Let's look at each of these parts of this verse individually. This verse of scripture makes it clear that man was God's ultimate intention when He created the universe. God started with a plan for His creative work and then followed that plan that culminated in forming man (and woman) out of the dust of the ground in God's own image.

It is very important to make the distinction that man was created as one man and woman as opposed to the fact that all the other living creatures were created in hosts that would cover the whole earth, not be located in one spot only. This would be necessary to keep the vegetation eaten back until man could populate the earth.

God, creating just one man and woman, would make it clear that, in the future, whatever directives God applied to Adam and Eve would apply to each of their descendants on an individual basis. Paul makes the point that when Adam sinned, all sinned who came after him. But the same is true of Christ as the new creation. Whatever applies to Christ as the beginning of the new creation, applies to all who are born again in Christ through faith. Because Christ is righteous, all those in Christ are righteous as well.

Jesus also refers to this same principle when He was asked about divorce in Matthew 19:3. He said God created one man for one woman and that was to be the pattern God expected for the rest of Adam and Eve's descendants to follow. But more than these reasons, God, by creating one man and woman, was establishing the fact that He desired to have a relationship with each individual who came into existence as descendants of Adam and Eve.

By telling them to be fruitful and multiply and fill the earth, God was saying, "I have plenty of love to go around for all those whom I bring into existence and the world I have created has great bounty to meet all their needs." That is true even today. God has plenty of love for all. That is why His word says, "whosoever will may come." There is also plenty of bounty for all to go around, but human greed is what keeps it from getting to everyone. I like to think of God's command to be fruitful and multiply like this.

Before God even created, He looked ahead and saw every person who would be born. He knew that many who would be born would reject Him and live to satisfy their desire to be their own god. But, on the other hand, He also saw all the individual souls that would receive God's love and live in His grace through His son Jesus Christ.

If you are a believer in Jesus Christ, when God said to Adam and Eve, "Be fruitful and multiply," He had you in mind. God was looking forward to the relationship He was going to establish with you.

As man fills the earth, God tells him to subdue the earth as well. This word "subdue" has the idea of bringing it under control. It was a challenge to conquer the earth. God was actually giving man the directive to study the creation; learn how it works; and use what he learns to become more and more effective in bringing the creation under his control for the good of mankind.

Before the fall, this would apply to working in the garden and expanding it as the human race began to grow. After the fall, man would learn that God created the earth in such a way that he could learn to improve his life in a fallen creation by increasing the study of science to improve medicines, increase productivity of land, and develop tools that would make life easier. In other words, this was a mandate by God to man to develop the scientific process. The mandate still applies.

Science in itself is not bad. Actually, Psalms 94:10 tells us God gives us all true knowledge, including scientific revelation. George Washington Carver admitted that his scientific discoveries were revelations from God in answer to his prayers for understanding. Most early scientist were God fearing men, such as Isaac Newton, Johannes Kepler, and Louis Pasteur, to name a few.

Science is not the problem. It is man, in his pride and rebellion against God, who determines to depend more on his science than on the Creator, who made science possible in the first place. Science, as we know it today, was begun by those who believed in God and used science to learn more about God through His creation.

Not only was man directed to subdue the earth, he was also given dominion or rule over all living things. This dominion suggested that man was created above the animals and that the animals were created for man's use in his quest to fill the earth and to subdue it. The animals were to be available to man's beck and call. We need to remember that this dominion was given before the fall. As we will see later, the animals, no matter what their size, were created in harmony with man and had no fear of man.

None of the animals ate meat and the fear of man was not given to them until after the flood. Man was a highly intelligent being and spiritual in nature. Chapter three (Eve and the serpent) tells us that man had the capacity to communicate with at least some of the animals before the fall, suggesting a possible telepathic or some verbal ability that was lost when Adam and Eve sinned. The Biblical account makes it clear that man had a special relationship with the animals and a power over the animals that still exists but is greatly diminished.

The command to fill the earth, subdue it, and to have dominion over all the animals makes it clear that God's original intent was to create everything and then give it as a gift to the ones it was created for, mankind. God originally determined to create man. Before He did, He went through the process of the creation week. He planned to give the creation as a gift to man. The creation would then provide an environment where man and God could develop a relationship with one another. That relationship is hinted at in Chapter 2 and suggested in chapter 3 when God came looking for Adam and Eve in the cool of the evening.

We also see this idea of relationship developed in the Old Testament stories of God relating to many different people such as Abraham, Moses, king David, and the prophets.

Actually, James 2:24 tells us Abraham was referred to as the "friend of God;" and Exodus 33:11 tells us Moses spoke to God, "face to face, just as a man speaks to his friend." God coming in the person of Jesus Christ to remove the barrier of sin also points to God's desire for our relationship with Him to be restored.

In chapter 1:29, God tells Adam and Eve that all the plants were to be food for them as well as all of the fruit from the fruit trees. Man was originally created to be a vegetarian. The plants were formed out of the elements in the ground and man's physical body was formed out of the elements of the ground. The vegetation would take the energy from the sun and use that energy to bring the nutrients out of the ground. Those nutrients would be transformed by the plants into food for man to eat. Later those same plants would also become the source for medicine to help man over come pain and suffering caused by the consequences of sin. I read a statement one time that I agree with.

It went something like this, "there is not a disease that man and animals have to contend with that God has not made a remedy for." It was just like a loving God to make provisions for the consequences of man's sin both physically in the created earth and spiritually through His Son, Jesus Christ. From the beginning, God had everything covered. In verse 30, we see that God gave the plants as food for all the living creatures to eat as well. None of them were to be meat eaters. That means that such animals as tyrannosaurus rex, lions, bears, wolves, etc. all ate plants. As the point has already been made, this fact makes it clear that God did not intend for death to be a part of His original creation.

Death is not good because death has no part in Who God is because He is eternal and does not experience death. God is the God of life, not death.

We are also told in Isaiah 11:6-9 and Isaiah 65: 24-25 that during the millennial reign of Christ all animals will eat vegetation once again. This indicates that the earth will be restored to the pre-flood environment during the millennium.

I personally believe that animals did not become carnivorous until after the flood because it is not until Genesis 9 that God is recorded as having made that change. There may have been scavengers that ate animals after they died, but there were no meat hunters before the flood. We will see later that, after the fall, the plant kingdom was changed in relation to man and, after the flood, the animal kingdom was changed in relation to man. Each change removed us further from the original creation making it impossible for us to conceive of how the pre-fall world really was. This is one reason why many cannot accept the Genesis account of creation. The world today is nothing like it was then.

We have no way of knowing how it was before the fall and the curses that followed. The present is not the key to knowing the past. Only God's word is. That brings us to the end of the sixth day of creation and the final pronouncement that all that God had made was "very good." This pronouncement tells us the finished creation was exactly what God envisioned it would be.

It was a perfect expression of the kind of God He is. He is not a god who created over ions of time using pain and suffering, death and dying. He is the God who calls all things out of nothing into existence immediately according to His will. He is the God who made all things in perfect order with exquisite design and purpose without survival of the fittest competition or laws of tooth and claw. It was a perfect paradise that would sustain life for as long as God willed and man lived in prefect trust of God. It also meant He was totally and completely satisfied with the finished product, especially with the creation of man.

The man and the woman were created in His image and now ready to take charge of this gift, the very purpose for God's creative activity.

Now it was time to take a break and begin to enjoy what He had made. The man and woman would join Him in His rest and celebration. They could appreciate the creation most because of their ability to relate to God in ways no other being could. Because they were the ones for whom He had created it all, they were to give all glory to God. In doing so, God would allow them and their offspring to enjoy His presence and love forever.

GENESIS CHAPTER TWO

We now come to the beginning of Genesis 2. Actually, Genesis 2 should not begin until verse 5. Verse 4 is where the first toledoth ends.

Understanding each toledoth refers to the information which precedes it, this toledoth in verse 4 would be the final summary of the creation account. This toledoth says, "This is the account of the heavens and the earth when they were created, in the day that the Lord God made earth and heaven."

So, verse 1 reemphasized the fact that God created every part of the universe, including the angels, stars, earth, host of plants and animals. Man is not included in this summary of all the hosts because the verse is directed toward man to remind him not to worship the creation or any part of it. Man was to worship only God Who created all things that exist because of him. This verse is an affirmation of the monotheistic worship of the one, true, transcendent, Creator God. All other worship is false worship.

LAW OF CONSERVATION

Genesis 2:2 once again reaffirms that God did all of His creative work in six days when it says, "And by the seventh day God completed His work which He had done..." Verse 1 and 2 both affirm one of the most proven laws of science: the first law of thermodynamics. This law is called the law of energy conservation. This law suggests that nothing new is being added to the universe. What is there has been there from the time of its coming into existence. It may change from matter to energy but nothing new is ever formed from nothing.

This law flies in the face of evolution, which suggests new forms and arrangements of matter are still coming into existence. An example of this new matter idea was the "Steady State Theory," that suggested as stars and galaxies moved apart from each other, other stars would form to fill in the gaps to maintain an equilibrium between the stars.

There is also a theory that suggests there is a place beyond the sun that continues to produce new comets. If the solar system is billions of years old, all comets should be gone by now, and yet, they are prevalent and still very pronounced in their structure.

This comet theory is another example of how the creation model predicts comets would still exist in a young universe whereas evolution has to contradict a law of science (conservation) to explain observable phenomena.

The important thing to understand about the law of conservation is science can only study what God is now preserving. God is no longer doing any creative activity. There is no way we can discover how God went about His creative work because He is no longer creating. We can only know the results of God's creative work, not how He went about doing it specifically.

The creation account tells us God spoke the universe into existence but it does not give us the specifics on how He formed the elements into matter out of particles that we still cannot comprehend.

THE GRACE PRINCIPLE

It was stated twice that God had finished His creative work in six days to put an added emphasis on that fact that God established the seventh day as a day of rest to complete the creation week cycle. The seventh day was to be a holy day to commemorate the fact that God finished His creative work in six days and rested on the seventh. Thus, God established for man the pattern of working six days and resting on the seventh day. Jesus tells us in Mark 2:27, "The Sabbath was made for man, and not man for the Sabbath."

The Bible also says that God does not slumber or sleep. He is not like man who runs out of steam and needs rest. The seventh day tells us that the first thing God wanted man to learn to do was rest. It also tells us that God made us for rest with Him. Notice, the first thing man had to do with God was rest.

It reminds us of Jesus calling out in Matthew 11:28, "Come to Me, all who are weary and heavy-laden, and I will give you rest." Jesus, as the Lord of the Sabbath, is the only one who could ever fulfill its meaning for lost man. When we put our faith in Jesus, we can rest in His finished work of salvation. We no longer keep the Sabbath as Christians because we keep the Sabbath through the rest we have found in Jesus Christ. The Sabbath is a part of the Law that Jesus told us He came to fulfill or bring to completion by His death on the cross. Matthew 5:17 makes that point, "Do not think that I came to abolish the Law and the Prophets; I did not come to abolish but to fulfill."

I cannot help but think that when Paul went into Arabia to re-study the scriptures in light of his new found understanding of Jesus Christ as the Son of God, that this passage was where he learned the doctrine of grace. Think of it, God did all the work of creation before He created Adam and Eve. When there was nothing left to create, then God created them. It is a strong statement that only God can create and man can only receive the benefits of His creation as a free gift.

There is nothing man could ever do to help God create anything. So God created the whole universe, turned it over to man as a free gift and then the first thing they did together was rest. Adam and Eve could do nothing to earn it or deserve it, they could only receive it as a free gift. There is a parallel here. Only God could do the work of salvation. When Jesus came to earth as a man, lived a sinless life, and was finally crucified on the cross, His final words were, "it is finished." This parallels God's word that pronounced His creative work finished in Genesis 2:1, "Thus the heavens and the earth were completed..."

Just as God did for Adam and Eve in giving them the creation as a free gift, Jesus did all the salvation work in our behalf out of His righteousness. What we could never do for ourselves (satisfy the law of sin and death), Jesus did for us and now offers to us forgiveness as a free gift. There is nothing we can do to earn it or deserve it, and so, He made it a free gift. That is a perfect expression of grace. God gives to us that which we do not deserve. The creation story and the salvation story both tell us that God is a God of grace. God wants us to enjoy His finished work of creation with Him. He wants us to enjoy His finished work of salvation with Him as well.

SEVENTH DAY FACT

Genesis 2:1-2 both make it clear that God finished His creation work in six days and rested on the seventh day. It seems our understanding of this fact that He created in six literal days was so important to God that He re-emphasizes the point in the Ten Commandments. Exodus 20:11 records for us the very words God Himself spoke audibly to the Israelites from mount Sinai. God said, "For in six days the Lord made the heavens and the earth, the sea and all that is in them, and rested on the seventh day: therefore the Lord blessed the Sabbath day and made it holy."

First, notice that the word day is preceded by the number seven both in this passage and in the Genesis 2:2 verse that says, "And by the seventh day God completed His work..." I bring this up to remind you once again that, in the Hebrew, the word day always means a literal day when attached to a number.

This is important because progressive creationists try to make the point that the seventh day is still going on as the period of time we are now living in. They try to use the word "day" in verse 5 as their proof because it obviously refers to the whole creation week rather than a literal day. But, we understand by the context that the writer is referring to a week not a long period of time. The seventh day was a literal day, just like the other six days as God tells us in Exodus 20:11.

There is a fact about the seven day week I want to reemphasize. It is a fact that often gets overlooked. We derive the length of a day by the earth's rotation. It takes twenty four hours for the earth to turn one time. We understand the length of a month on the basis of the moon in orbit around the earth. It works out to thirty or thirty one days except February.

The length of a year is determined by how long it takes for the earth to orbit the sun one time. It takes three hundred sixty five days. The question is, " How did we decide to break our months and years into seven day weeks?" The answer is, there is no physical reason for establishing a week as seven days.

Genesis is the only place where we find the establishment of a seven day week in all of antiquity, and yet, all calendars use a seven day week. That would make for a very strong suggestion that the Genesis account of a six day creation week with a seventh day rest is the origination of the practice from the beginning of time. After the flood, when all the people were dispersed at the time of the tower of Babel, they continued to use a seven day week as is suggested by the fact that the earliest calendars used a seven day week. It was a practice so firmly entrenched before the flood that it continued after the flood being practiced by all the families dispersed from Babel.

CHAPTER 9
WHICH ACCOUNT?

This chapter is titled, "Which Account", because many try to say that the account given in Genesis 2:5 and following to chapter 3 is a different creation account then the one found in Genesis 1:1-2:4. They use arguments like a new name for God, a different chronology of creation events, and the different creation of Adam and Eve. Usually those who give those arguments come from the Documentary Hypotheses school of J,E,P,D, and other forms of Higher Criticism. This is a system of interpretation that has its roots in evolution theory. Like Hegel's evolution based theory on the development of history, this German school of Biblical interpretation rationalized that scripture was a product of an evolving development in Hebrew culture and religious thought as an off-shoot of early Sumerian and Babylonian polytheism.

They supposed that early documents were pieced together by men who wanted to create a centralized basis for their faith as Israelites. They rejected the idea that Genesis was compiled by one man, Moses, as a historical account of Israel's history from creation to the Exodus. The literal creationist, on the other hand, sees this account in chapter 2 as an embellishment of the events of the sixth day as witnessed and recorded by Adam himself. The account then, does not contradict Genesis 1 but continues the revelation by God's first inspired recorder of His word, Adam.

Like in all acceptable practices of Biblical interpretation, scripture is its own best source of understanding scripture. This being the case, Genesis 1 becomes the authority for understanding Genesis 2. What is written in Genesis 2 should be explained by what has already been stated in Genesis 1. When you take this approach, it all fits into a logical explanation rather then creating confusion.

THE AUTHOR

Who wrote Genesis 2? That is the major question. The most reliable answer should come from the text itself. First, if this is a historical account, then it would seem reasonable to assume that Adam himself would be the only eye witness to these events besides God. Secondly, we find Genesis 5:1, the second toledoth, tells us that Adam is the source of the account written from Genesis 2:5 to Genesis 5:1. The toledoth says, "This is the book of the generations of Adam." Notice the word "book." This verse suggests Adam wrote the book or was the source for the second toledoth. He may possibly have dictated it to someone much like Paul dictated his letters to others while he was in prison, but Adam is stated clearly as the source. It also suggests to us that Adam had the capacity to write. This is not hard to believe when we realize he was probably the most intelligent man (made in the image of God) who ever lived. This man had the ability to communicate with God from the moment he was created. This is evidenced by God giving Adam instructions about the two trees in the garden on the same day he was created. We have already seen that God wrote the Ten Commandments with His own hand and gave them to Moses. Surely God could teach Adam to write, as well as, how to take care of the garden, as we will see later.

Third, the context itself suggests Moses, as the compiler of Genesis, would have added his own commentary to help the people of his day remember what the world was like before the flood. Moses wrote over eight hundred years after the flood to the Israelites after they came out of Egypt. One example of Moses' commentary would be reminding the Israelites in Genesis 2:6 that the ground was watered by a mist because there was no rain before the flood. In Moses' time, there was rain so his readers would benefit from this information.

If we let the scripture's worldview be its own interpretation, then Adam makes perfect sense to be seen as the author of Genesis 2:5 through Genesis 5:1. Adam was not a product of the caveman to modern man evolution process. Adam was created totally capable of talking, learning to read and write, knowing how to take care of the garden, and keep a calendar. He certainly had the ability to understand where the earth was in relationship with the sun, moon, and stars in the process of a year. Remember, the sun, moon and stars were created for times and for seasons and Adam knew how to read this astronomical clock. The ages of the men , including Adam, in Genesis 5 tell us Adam, from the beginning, knew how to count the years. He was highly intelligent and so this would not have been a challenge for him. Only evolution theory would lead us to believe any differently. Adam, being the author, allows us to see his account as an eye witness inspired record passed down to those generations that came after him. His account then, became the first addition to God's progressive revelation of Himself through the written word. Every generation after Adam had the creation account and Adam's testimony to guide them in their understanding and worship of the one true God. We will also see later that this account was written after the fall.

The structure, the content, and the temperament all suggest Adam wrote this account to explain why the world and men were so evil after the fall, when God was supposed to be a good God. Adam confesses by his own testimony that the fall was a consequence of his own rebellion against God. Adam admits he was the problem not our good God Who made all things good.

THE TRANSITION

Genesis 2:5-7 gives us transition information before we come to events that Adam would have witnessed first hand. We must keep in mind as we study these verses that Genesis 1:1-2:4 is the foundation for understanding whatever comes after it. That first toledoth is the original creation story that guides us in understanding any other information that comes after it that may seem to contradict what has already been recorded. This is especially true of what is recorded in Genesis 2:5-25.

In Genesis 2:5 we are told, " Now no shrub of the field was yet in the earth, and no plant of the field had yet sprouted, for the Lord God had not sent rain upon the earth; and there was no man to cultivate the ground." The first apparent contradiction comes from the statement about the "shrub" and the "plant." Genesis 1 tells us God created all vegetation on the third day so this verse, at first, seems to suggest God's creation of shrubs and plants was not complete on the third day. The important words to consider here are the modifiers of shrub and plant which are, "of the field." This verse is not telling us the shrub and plant did not exist; it is telling us that these kinds of vegetation had not yet been cultivated into organized fields.

Cultivation would not be necessary until after the fall, when man and woman were forced out of the garden and cursed to till the ground by the sweat of man's brow. Before the fall, Adam and Eve lived in a garden amongst the fruit trees and did not rely on cultivation of fields.

Verse 5 then adds the reminder that in the pre-flood environment there was no rain, which would be consistent with a canopy covered environment. The man was not yet created and so this would be a statement of how God saw His creation after the forming of the land animals in preparation of creating man. In other words, the creation was incomplete without man. The verse also seems to be prophetic in that it suggests man would learn to cultivate the earth and bring it all under his control in time. This would imply that Adam's readers of his account were now living in agrarian cultures after the fall and needed this clarification.

Verse 6, "But a mist used to rise from the earth and water the whole surface of the ground," then explains how the earth was watered before the introduction of rain at the flood. As I said earlier, this would be commentary added probably by Moses to the readers of his time after the flood. It does tell us that there was a very different system for watering the earth before the flood. This system had to have been much more efficient than the rain cycle we have now. The water would come from the saturated ground containing the fountains of the great deep and surely there was dew that formed from humidity in the atmosphere close to the ground. This would suggest an earth that was saturated with plenty of water implying no deserts existed in the pre-flood world and no droughts.

The vegetation would thrive in a near hydroponic environment that would make it possible for plants to grow fruit and vegetables to enormous size year round. Verse 7 is key because it gives us more specific information on how God actually formed man out of the dust of the ground and would then form the woman from matter taken from the side of Adam. It tells us God took a hands on approach when He created Adam and Eve, something that is not stated about the angels or animals. It also tells us that only man received the breath of life or spirit of life, making man a very distinct part of God's creative work. Verse 7 says, "Then the Lord God formed man of dust from the ground, and breathed into his nostrils the breath (spirit) of life; and man became a living being." Not only was man a living being , his life or *nephesh* was empowered and embellished by the very Spirit of God making man a body, soul, and spirit being.

This fact separates man from animals in that animals have created soul life but not God breathed spirit soul life. Zechariah 12:1 emphasizes the creation of man as having spirit life when it says, "Thus declares the Lord who stretches out the heavens, lays the foundation of the earth, and forms the spirit of man within him..." Man, embodying the life of God, makes him an eternal being whereas animals only have created life but not spirit life and, thus, are not eternal. When an animal dies, it ceases to exist. We know there will be animals created for the new heaven and earth because we see Jesus returning to earth on a horse, and animals were pronounced good as a part of the first creation. We would not know these important elements of man's creation without this 7th verse.

There is an important point I must interject here.

The book of Job is considered by many to be the oldest book in the Bible, besides the pre-Abraham records. Job takes place after the flood and several hundred years before Moses compiled Genesis. There are many important verses in Job that make it clear that the information in early Genesis was passed on to those in the Job story long before the time of Moses. This supports the premise of this book that the accounts of the creation, the fall, and the flood were passed down to post-flood man either by written records or word of mouth. I have no problem accepting the written record position. Job 33:4 is one of those verses that support the theory that post-flood man had the pre-flood records in hand. The verse quotes Elihu as saying, "The Spirit of God has made me, and the breath of the Almighty gives me life." It is more than reasonable to assume that Elihu understood his own life was a continuation of what God started here in Genesis 2:7. Obviously, Elihu had the information given in Genesis 2:7 available to him after the flood, probably in written form.

What a beautiful picture. Think of it. God lovingly, with personal hands on care, formed Adam's body with great anticipation of the moment when Adam would become a reality. God had done all of His work in preparation for this moment when God would finally breath His very own Spirit into Adam's nostrils. And at that moment, as Adam received that breath of God's life, he all of a sudden became conscious as that life filled his perfectly formed body. He opened his eyes and the first thing he saw was the glorious face of God, his Creator, pulling away from him as He had just finished breathing into his nostrils. I cannot help but believe that God had a smile of satisfaction on His face as He took Adam's hand and helped him to his feet.

What a blessed and glorious moment that must have been. Adam may have asked, "Who am I, and where did I come from?" He may have also asked God, "Who are You and where did You come from?" Genesis 2:7 (the creation of man) answers the first two questions. The creation story in Genesis 1 answers the second two questions. Not only did they answer the questions for Adam, they also answer the same questions for us. All we have to do is trust God's word just as Adam did. Philosophy in light of God's word is really that simple.

In Genesis 2:4 there is a new name used for God. Now God is referred to as "Lord God" or Jehovah *Elohim*. In all of the creation account from Genesis 1:1 to Genesis 2:3, God is translated from the Hebrew, *Elohim*. This strongly suggests a new writer is now writing. It seems plausible to assume that God, in His own account of His creative activity, would refer to Himself only as *Elohim*. But, when Adam wrote his continuing story, he would refer to God as, Lord God, especially if he wrote after the fall and no longer had direct access to God due to his fallen state.

The title, "Lord God," would express Adam's sensitivity to God's holiness and Adam's own foolish choice to become like God that led to his fall. "Lord God," being included in the first toledoth recorded in Genesis 2:4 would suggest Moses, as the compiler of all the eye witness accounts handed down to him, would add the *toledah* to end each account to make a distinction between them. This would explain why "Lord God" is used to finish God's own record of His work of creation. Moses would use the same reference to God as Adam because Moses also lived in a fallen state before God.

THE GARDEN

The creation of Adam is now complete. The question is, where is he going to stay? God already had the answer in mind. Instead of putting Adam in a forest jungle to fend for himself, God, in His grace and love, prepared a special place for him. This planting of the garden reminds us that God is preparing a new heaven and earth for those to dwell in who chose to receive God's love and grace as shown in Revelation 21 and 22. It reminds me of what Jesus said in John 14:3, "I go to prepare a place for you." Here we see God preparing a special place for Adam and Eve, who was soon to come into existence as well. Verse 8 says, "And the Lord God planted a garden east in Eden; and there He placed the man whom He had formed."

It is important to note that God "planted" this garden. This would mean that God used already created seeds from the third day of creation rather than doing a new creative act. A new act would contradict the original creation account, but God planting already created seeds agrees with Genesis 1:11, "Then God said, "Let the earth sprout vegetation, plants yielding seeds, fruit trees bearing fruit after their kind, with seed in them, on the earth," and it was so." God took what was already created and used it to plant the garden. This planting of the garden after the creation of the man also tells us Adam was present at this event on the sixth day. This makes it clear that God used this as an opportunity to show Adam how seeds and plants work. Adam would need this information to become the caretaker of the garden and know how to expand it in the future. This would be a crash course since God had other things to do with man before the sixth day was done. Man was certainly smart enough to handle the new information.

This tree planting and maturing in a part of a day means the process took place immediately. This planting of the garden reminds us of the miracle of Jesus when He caused the fig tree to whither immediately. Here we see that God would have no problem causing the seeds to grow immediately into mature fruit trees in the opposite way of causing them to wither.

Genesis 2:8 also names the garden, "Eden" (meaning delight), and gives its location as being toward the East. This would suggest that Adam was writing to his descendants who would understand where Eden was from their own locations on the earth. This also suggests that Adam wrote to them several years after leaving the garden. It would be much like today. When we speak of the East, we refer to the area of the earth that includes the Bible lands even though they are south of Russia and west of Japan. Adam is giving information to people who may have been somewhat removed from knowing the events and locations Adam now records. We know it would not benefit Moses' readers because Eden would have been totally buried by the flood and nowhere to be found in Moses' day. This is Adam giving information or clarification to readers who have come after him and are now dispersed on the earth. Adam does not want them to forget where the garden was so they could still find it if they looked for it.

A good question to ask is who named the garden, "Delight," or at least the area where the garden was located? Was it God or was it Adam? We are told later in this chapter that God brought the animals to Adam to name. In Genesis 3:20, we are told that Adam named his wife, Eve, because she was the mother of all those to come after her.

God did not really name Adam because the word "man" in the Hebrew is *adam* suggesting Adam was not originally a name. There are several suggestions as to how *adam* is to be literally translated but it most likely means, "made of the earth."[32] The name, "Adam," then was derived from God's description of the man rather than God giving him a specific name. My point is that the name, "Eden," most likely came from Adam himself and is a documentary on Adam and Eve's time spent in the garden with God. That time spent with God before they fell into sin and were separated from God and the garden. It was a wonderful experience of love, joy, and peace; a place of delight. Their Eden experience totally contrasted the world Adam, Eve, and their descendants had to live in outside the garden. This world was a place that was ruled by God's curse of thorns, thistles, hard work, pain, suffering, and death. The name "Delight" was a statement of paradise lost. At the same time, it also reminds us of the hope we have in the promised new heaven and earth that God intended for us to enjoy all along. Our life with God, in the new heaven and earth, will once again be a delight.

Genesis 2:9 gives further information about the trees that God planted in the garden. There were trees of all kinds that bore delicious fruits that were pleasing to the eye and good for nutrition. In the middle of the garden, God planted a tree called the "tree of life;" and He planted another tree called "the tree of the knowledge of good and evil." In the context of this verse we can conclude that these two trees were planted trees just like the others. They were not magical trees. This is important to understand when we are making the point that this is not a mythological story but is a historical eye witness account.

These two trees were just as much a part of the creation and just as physical as the rest of the trees that were all made out of the same elements. The tree of life may have had wonderful capacities for healing but that does not mean it had more than just physical qualities to it. Almost all the medicines we have today come from plants. Those medicines have the ability to bring healing and lengthening of life just as the tree of life would have had only in a much stronger way.

The tree of the knowledge of good and evil did not have to be extra ordinary either. It was man's sin of disobedience to God that caused him to fall into sin, not any special qualities of the tree. As a matter of fact, Satan, the liar, tries to convince Eve that it was a magical tree by telling her it had the power to make a person wise. Wisdom comes from learning and following the ways of God, not eating from some magical fruit. This will be discussed more fully later. Suffice it to say at this time, it was the prohibition of God to Adam not to eat of the tree that made it special, not any special ability to give some mystical knowledge. There is no such thing as mystical knowledge. The only true knowledge that leads to wisdom and life is the knowledge we learn from God by reading His word and applying it to our lives.

The next thing we learn about the garden in verses 10-14 is that a huge river flowed out of the garden and this river helped to water the garden. We know it was huge because it fed into four other apparently large rivers. Just as Adam reminded his readers that Eden was in the East, he now gives more specific information concerning the garden's location by describing these four rivers and what the land was like that they flowed through. He especially mentions the land of Havilah where there was gold, bdellium, the onyx, and the land is good.

The mention of these metals and stones tells us his readers had already developed a value system apparently for trade purposes. The mention of the good land would suggest a place that was good for cultivation and the establishment of culture. This is consistent of what we will read about the development of technology in chapter 4 where we will see the descendants of Cain developing culture. All this means that Adam was writing after the fall and the human race had already developed culturally to a degree.

Two of the river's names, Tigris and Euphrates, are of interest because they are found after the flood as well as here before the flood. They also travel through the land of Assyria, which is also named after the flood. Some try to use these names to argue that this account was written after the flood by those who were acquainted with these post-flood names and places. The better explanation is that Noah and his family named these two post flood rivers after they left the ark on Mount Ararat and migrated into the Mesopotamian valley. When they encountered the two rivers in the valley they named them after the rivers they knew before the flood. The land they flowed through were once again named Assyria as a reminder of the pre-flood world. It was much like those who traveled from England and Europe to the Americas. They named towns and rivers after those they remembered in the old land. These rivers in Genesis 2 were totally destroyed in the flood. Only the names survived in the minds of Noah and his family. The fact that the Pishon and the Gihon, also mentioned in Genesis 2, are not found in close proximity to the Tigris and Euphrates with a common headwater, and the name Havilah is used after the flood with no connection to these rivers, makes it clear that this area of land described in Genesis 2 no longer exists.

It was all buried in the flood and only remained in the memory of Noah and his family.

THE COMMAND

Adam's account now moves from giving explanations about the garden to setting the stage for the explanation of why his descendants could no longer live in the garden of delight. Man's life in the garden with God was going to have conditions and consequences. In verse 15, it is made clear that Adam did not have anything to do with the planting of the garden. This is still the sixth day of creation and God is doing all the work. This being true, verse 15 says, "Then the Lord God took the man and put him in the garden of Eden to cultivate it and keep it." Here we see that Adam was to cultivate the garden of trees, but had no part in planting it. As God put Adam in the garden, this would be the time God showed Adam how it all worked. Later, outside of the garden, Adam would cultivate the land with shrubs and plants using what he learned from his work in the garden. Keeping the garden would be Adam's job, but his work would not begin until after he spends a day of rest with God. The work Adam would do in the garden would be nothing like the toil of blood, sweat, and tears he would be doing after the fall. At this time before the fall, all things were in perfect harmony and work for Adam was as he called it himself, a delight rather than toil.

As part of God's instructions to Adam on how to cultivate the garden, God brings him to the middle of the garden where the two important trees have been planted by God. Now God's instructions turn into a solemn command.

This was what God said, "...From any tree of the garden you may eat freely; but from the tree of the knowledge of good and evil you shall not eat, for in the day that you eat from it you shall surely die," (Genesis 2:16-17). This was a strong command with a very severe consequence if disobeyed. Up to now, everything about creation was about life.

Here we have the first mention of death. That tells us that as long as Adam obeyed God in this one command, death would not be a part of Adam's experience. He could live with God forever. Adam's testimony here denies the possibility of the evolution process of living and dying organisms competing with each other for supremacy and survival. One or the other is true but they both cannot be true. An interesting thought is that, when Adam was without sin, he only had to deal with one command. He needed no others because he was totally good and did everything out of that goodness naturally. Now that command has turned to ten commands as well as many others to keep man's fallen nature under control. The lesson is the more we disobey God, the more freedom we lose.

CHOICE

Why did God include, in a sense, this temptation of eating from a forbidden tree as a part of His creation and then, declare it all very good? I believe the only good answer to that question is choice. But why choice? Why is choice so important that God would include it in His creative plan and call it good? He knew full well man would fail the test and, ultimately, the Son of God would have to be sacrificed to save mankind from their sins. I believe the answer is that to be made in the image of God man had to be a fre moral agent.

Why? God Himself is a free moral agent. Man could not be free without having the ability to choose to obey God or to disobey God. Choice means God created us for freedom. Freedom requires choice and choice makes freedom possible. Think of it like this. The people in a nation ruled by a dictator are not allowed to choose their leader. Their leader has forced himself on the people.

He makes all the decisions whether the people agree with them or not. The dictator chooses all those who are a part of his government, not the people. God is telling Adam he can choose to let God be his God; or he can choose to be his own god. This responsibility of choice allowed their relationship to be based on freedom, not a dictatorship. The problem is, God created the universe as a totally moral and just universe. He could do nothing other then that because God is totally moral and just. For man to choose to be his own god would be totally immoral and unjust; lacking gratitude and humility all of which is not good. This being true, the consequences of such a choice to reject God and be one's own god had to be death.

This choice to disobey God would be a total violation of man's being made in God's image. God, being good as the Creator of the universe, required man to be good as a perfect expression of God's image. Man's rebellion against God would put man in complete opposition to the essence of the total fabric of goodness the creation was created in. This violation of the essence of the good creation required accountability.

Man was created to be good living in a good creation. Thus, when man did reject God, God had to curse the creation to fit the fallen qualities of man rather than the goodness of God.

If man was no longer good, then the creation could not be good any longer because the creation was created for man. It became a place for fallen man to dwell in as his own god, but no longer a place for the presence of a good God. We will see this in more detail in the record of the fall in Genesis 3. Choice is not just about freedom. Choice is about trust, love, responsibility, and accountability.

When God gave Adam this command to not eat of the forbidden tree, He was saying, "I want you to trust Me." Out of that trust, Adam had the opportunity to grow in a love relationship with God. Love is a product of trust. God was also saying to Adam, "I am willing to take the risk of giving you choice to show that I want to trust you knowing that, if you break that trust, it will cost Me the cross." This concept of trust applies to the exception clause that Jesus made to the law about divorce in Matthew 19:9. The exception was that divorce is only allowed when a spouse has been morally unfaithful. Why would this be so? Sexual intimacy is the most sacred part of the marriage bond. To be sexually unfaithful is to destroy the trust of one's own marriage partner. That is why the seventh commandment specifically says, "Thou shalt not commit adultery." Adultery applies specifically to being unfaithful to your marriage vows. Adultery destroys trust, and a relationship without trust undermines a person's ability to love. A relationship without love then, becomes a tasteless arrangement like chocolate pudding without any sugar. It is bland and not eatable. A marriage without love would be a curse, and God intended marriage to be a blessing. It can only be a blessing when there is trust in the relationship.

Trust makes real love possible; but choice also requires responsibility.

God giving Adam and Eve choice means that they were free moral agents who had soul competency. They had the capacity to know the difference between right and wrong and could be held responsible for their choice. Freedom that comes from choice is a huge responsibility. We see many people today who want to be free to do whatever they want, but then not have to take responsibility for the consequences of what they choose to do.

They want free sex but not the responsibility of taking care of a child resulting from that sex so they choose to abort the child violating the child's right to life. That is not freedom, that is anarchy because freedom has to take into consideration how our actions impact the freedom of those around us. We see that when Adam and Eve chose to disobey God, they were only thinking about what their choice meant to them at that moment. They thought nothing of what it might also mean to God and their relationship with Him, let alone their descendants coming after them. That is how it is with sin. Sin is usually momentary and not very visionary. If we have freedom, we are responsible with what we do with that freedom in the way we make our choices.

The command that God gave had a terrible consequence. Verse 17 says, "...for in the day that you eat from it you shall surely die." We are responsible for what we do with our freedom, and this verse tells us that responsibility requires accountability. In other words, God is telling Adam that your choices have consequences. If Adam chose to trust God, he would enjoy all the wonder and delight of God's creation. Not only that, he would also enjoy the very presence of God and fellowship with God for eternity. On the other hand, if Adam chose to distrust God and eat of the forbidden tree, then all these benefits would be forfeited.

From our fallen position, we often think that this seems like an awfully severe judgment but then, we have no clue as to what was at stake from God's perspective. We also have no real appreciation for the complete and total holiness of God. But, we must admit that God was totally and completely just in following through with His warning because He did let Adam know the rules from the beginning. God, in His own justice, was duty bound to Himself and Adam to stand by His word.

Thank God that He not only stands by His justice, but also stands by His promises to forgive and heal all those who repent and turn back to Him through His son, Jesus Christ. But remember, choice is about freedom, trust, love, responsibility, and accountability. You cannot have a meaningful relationship without these ingredients.

In His warning of the consequence of death, God says. "...you shall surely die." There are those who try to use this as another contradiction in the Bible. They point to the fall in Chapter 3 and say that, after Adam and Eve sinned, they did not die just as Satan said they would not die. Who are we to believe, they infer? Another way of translating this statement is, " In dying you shall die." This translation tells us God was saying that Adam would experience a death that would lead ultimately to a total death. Remember, Adam was alive in body, soul, and spirit. As we will see later, this verse helps us to understand what took place when Adam and Eve ate of the tree of the knowledge of good and evil and their eyes were opened and saw that they were physically naked. These verses and others that Jesus gave later make it clear that Adam and Eve experienced a spiritual death immediately and began to die physically from that moment on. We will speak more of this in chapter 3.

THE WOMAN

Now that God and Adam had come to an understanding about their covenant together about choice, God was going to finish His creative work by giving Adam the greatest gift of all, a partner, a friend, a helper, a soul mate, a wife. Verse 18 is a transitional verse that leads into this final creative act of God. It says, "Then the Lord God said, 'It is not good for the man to be alone; I will make him a helper suitable for him.'" This verse tells us that the rest of the chapter is to be understood in the context of God providing Adam with Eve. This is especially true of the naming of the animals.

But before we get to the naming of the animals, we need to see the principles that are indirectly stated in verse 18. First, we see the use of the words "not good." This "not good" suggests to us that the creation would not be complete without Adam and Eve being created together in perfect harmony enjoying relationship. It was not good for either to be alone. Quality of life includes relationships. Those who try to say that all they need is God do not have God in agreement with them. We need people as well as God in our lives. Adam and Eve together, as we have already said, were the expression of the image of God. People needing people then became good. Secondly, if it was not good for Adam to be alone, then it would not be good for the woman to be alone either because she was made to be Adam's companion. That being the case, then Adam was made to be Eve's companion as well. There are those who were not intended by God to marry, but it seems that those who do marry and have children have the opportunity to have a fuller and richer life together then they would if they did not marry one another.

Being married and having children is a complete life that provides opportunities for growth and understanding not directly accessible to those who remain single. The key is being sure to find the right partner that God has prepared for us. Third, Eve was to be a helper suitable for Adam. Actually, they were suitable for each other. This brings up the principle of servanthood. None of us were made to be served but to serve. Eve was capable of meeting all of Adam's needs and Adam was capable of meeting all of Eve's needs. Of course those needs did not include those needs that only God Himself can meet in our inner most being. A truly quality life is living to meet the needs of each other out of a meaningful relationship already established with God by each individual.

Why is the task of naming the animals included in Adam's account of God's making of Eve? It is another one of those passages that scholars have tried to use as a proof that Genesis 1 and Genesis 2 are two different accounts of creation. At first glance, it seems that verse 19 is saying that God created the animals out of the ground after He had made Adam. The verse says, "And out of the ground the Lord God formed every beast of the field and every bird of the sky..." Remember, the birds were made on the fifth day and the land animals on the first part of the sixth day. The land animals here are mentioned first and the birds after them. The sea creatures are not mentioned at all, but then how could God bring them to Adam to name without them all dying out of the water or Adam getting wet by going to them in the water?

How do we keep this verse from contradicting what Genesis 1 has already told us about the steps God took in creating the sea creatures, the birds, the land animals, and then man?

Recognizing that the Bible is its own worldview and thus its own best source of understanding, we must see what Adam already knew when he wrote this account. We must keep in mind that Adam had the Genesis 1 record available to him when he was writing down his experience. There is no contradiction to the fact that God had formed the birds and land animals out of the ground as verse 19 states. The fact that God "formed" (past tense) the animals can be understood to mean that God had already created these animals from the ground and was then bringing the animals to Adam to name. Apparently the animals were outside of the garden until God Himself brought them in for Adam to see and then name. If Genesis 1 is our authority on how to understand chapter 2, then this makes perfect sense. There is no real contradiction here. The only reason it would not make sense is if the person reading this passage was already committed to the interpretation that Genesis 1 and 2 were two different creation accounts.

Another argument skeptics use to question this account is the idea of man having to take the time to name all the different kinds of birds and land animals. It would take several days rather than hours they argue. But, that depends on what Adam would mean by all the animals. Adam would only have to name the kinds, not every subdivision of kinds, and he would only have to name the animals God brought to him. Remember, Adam did not name any sea creatures at this time, and the real reason for this task was to cause Adam to see that only Eve could meet his needs, not any of the animals. God would only have to bring enough animals to make this point clear. Obviously, there would have been many to name, but this naming of the animals also shows the high intelligence of Adam and the rapidity by which he was capable of completing this task.

It also affirms God's sovereignty over the animals and His total and complete ability to bring the animals to Noah to find shelter in the ark when it will be necessary at a later time. There are two basic reasons for God wanting Adam to name the animals. First, God wanted Adam to learn he is not an animal and that he had authority over them. His giving the animals names suggested Adam's position of superiority and ownership.

Second, God wanted Adam to see that there was a separation between him, as a human, and the animals, as animals. When Eve was brought to Adam, he would then understand the perfect complement she was to him and that she was to be the only source for having his needs met that was acceptable to God. This whole exercise establishes that bestiality and homosexuality are not acceptable to God and had no part in the original creation. It also makes it clear that God intended for one man committed to one woman for life to be the standard for all time. The last phrase of chapter 2:20 sums up these points by stating, "...but for Adam there was not found a helper for him." Eve was to fill all voids in Adam's life. God Himself created those voids by making Adam a being who was to be fulfilled by his relationship with Eve.

Now that Adam understood that his fulfillment was not going to be found in the animal kingdom, God had gotten him ready to receive this special gift He had in mind for Adam all along. But it was going to require some self-sacrifice on Adam's part in order to receive the gift. Adam had to undergo a removal of a part of himself, by surgery, to receive Eve. We also have to undergo some surgery of self to receive the joys of salvation. We have to repent of our pride and rebellion and put to death the old man in order for the new man to be able to grow and mature in us.

Chapter 2:21 and 22 tell us how God made Eve. He put Adam to sleep and then took a rib or a part of Adam's flesh from his side. The Hebrew is not clear about how to interpret this. Either way, the result is the same. God took matter from Adam's body to make Eve's body. Eve, being fashioned from Adam's side, tells us that she was made from the same elements as Adam. It also tells us that when God made Adam, He was also in the process of making Eve. From Genesis 1:27, where we are told that both Adam and Eve were made in the image of God, we understand that God would have breathed into Eve the same breath of life He breathed into Adam. Eve's experience of seeing the face of God for the first time was similar to Adam's. The whole process communicates to us that Adam and Eve were to understand that they were inner dependent and God solidified this point when He announced that the two were to become "one flesh" in Genesis 2:24.

Eve being taken from Adam's side makes a very important point. We are told in Genesis 3:15 that the Messiah was going to come from the seed of the woman but, because the woman came from Adam's body, then her seed originated in Adam himself. This would then mean that the Messiah coming from the woman's seed would still have a part in Adam. This being true, He could be the sacrifice for all the rest of us who were also in Adam when he sinned causing us all to be sinners as well. The big difference would be that Jesus received his physical body from his mother's seed that was quickened by the Holy Spirit as a new creation combining the physical body of a man with the sinless nature of God Himself. That is why Paul could refer to Jesus as the new Adam or the second Adam. He came from Adam and yet was God at the same time. We will look into this more fully in Chapter 3. There is a New Testament parallel that is prophetically pictured here. After Jesus had died on the cross, a Roman soldier took a spear and thrust it into Jesus' side causing blood and water to flow out of the heart cavity of Jesus. The church in the New Testament is often referred to as the bride or wife of Christ. Paul says in Ephesians 5:25, "Husbands, love your wives, just as Christ also loved the church and gave Himself up for her..." Just as Eve came from the wound of her husband's side, so the church, the bride of Christ, was established by the blood that flowed from the wound in Jesus' side. As the new Adam, Christ received his wife from the wound in His side just as the first Adam received his wife from a wound in his side.

After God "fashioned" the woman, He brought her to the man. I can just see this scene in my mind as I picture God finished with His crowning masterpiece of creative work and bringing her to Adam knowing full well what his response was going to be. Adam's response to this precious gift was exactly what God expected. The Hebrew wording suggests a cry of joy when Adam first saw Eve and responded by saying something like, "Wow! This is now bone of my bones and flesh of my flesh; She shall be called Woman because she was taken out of man." Remember, Adam had just named the animals, and now, with understanding and great appreciation, he pronounced this newly made gift, "woman." Because she was made from his own physical body, Adam received her as a part of himself, understanding himself to finally be a completely whole person.

The same would have been true of Eve. She was also made complete. Think of it, as God brought her to Adam, she would have been asking questions about him as God prepared her for their first introduction. God, with a grin of silent joy, probably did not say too much to Eve in reply knowing, that when she saw Adam, all her questions would be answered. God brought the two together, as He does for all marriages made in heaven, becoming the original Match Maker. It was a perfect match.

THE STANDARD

Jesus tells us in Matthew 19:5 that chapter 2:24 were the very words of God who inspired either Adam or Moses to write after this event, "For this cause a man shall leave his father and his mother, and shall cleave to his wife; and they shall become one flesh." God taking the woman from the man established by this historical act that marriage was to be between one man and one woman for life. Jesus also adds His further interpretation of this event by saying in Matthew 19:6, "Consequently they are no longer two, but one flesh. What therefore God has joined together, let no man separate." Jesus makes it clear in the rest of the Matthew 19 discourse that this standard had not changed, but the fall of man had distorted man's ability to live up to the standard.

In speaking of the standard, all the events described in Genesis 1 and 2 took place before the fall. God pronounced the way things were before the fall as the perfect expression of the kind of good God He is, thus establishing His standard for what is good. Life is good, and so murder is wrong. Choice is good and so tyranny is wrong. One man was made for one woman, and so fornication, adultery, homosexuality, bestiality, premarital sex, polygamy, prostitution, pornography and all other sexual acts outside of marriage are wrong. God made plenty of land and food for everyone, and so hoarding is wrong. God is good and so whatever is not good is wrong. All these and many other things are wrong because they violate the original standard of good or right God established before the fall.

The standard has not changed even though man, in his rebellion against God, has changed. God is the same yesterday, today, and forever. True followers of the Creator God revealed in Genesis 1 and 2 will try to live their lives according to God's standard. Those standards are the foundation of all morality, ethics, and law. The founding fathers of the United States referred to the standards as the "laws of Nature and of Nature's God." They define what is right and what is wrong based on the moral character of the good Creator Himself.

NAKEDNESS

The final verse in Genesis 2 makes an important point. Verse 25 says, "And the man and his wife were both naked and were not ashamed." What does that mean? First, our physical bodies in and of themselves are not sinful. Our sexuality was pronounced good as a part of God's creation. The sexual relationship between a husband and his wife in marriage was intended by God to be a blessing. Second, before the fall, Adam and Eve had received the "spirit" of life making them spiritual beings. They were able to have fellowship with God, Who is Spirit, in the garden. Their not being ashamed of their nakedness would suggest that they were living out of their spiritual existence and their physical bodies were instruments of their spirit.

Man was created to be a spirit, soul, and body tri-unity as an expression of the image of the triune God. Before the fall, man lived from his spirit, through his soul, directing his body. After the fall, it was reversed. As we will see later, man died spiritually and became controlled by his physical environment appealing to the senses and appetites of his flesh.

This became the rudimentary reason for all addictions which are expressions of the flesh ruling the soul. His soul (mind, will, and emotions) fell under the control of his flesh, which became out of control without the direction of the spirit. Being born again, as Jesus described the quickening of the spirit in John 3:3-8, is the return to the pre-fall state where the spirit once again can be set free to control the soul and body. A Christian is free to choose to live from the inside out once again. All this can be summarized by saying a lost person lives from the outside in; a born again Christian can learn to live from the inside out. That is what it means to grow in Christian maturity; living under the control of the spirit, not the flesh. The spirit is the new man that is to put the flesh or old man to death.

Paul refers to the three levels of man's being in I Thessalonians 5:23 when he writes, "Now may the God of peace Himself sanctify you entirely; and may your spirit and soul and body be preserved complete , without blame at the coming of our Lord Jesus Christ." Just as Adam and Eve were a spirit, soul, and body, every Christian who is born of the Spirit once again becomes a spirit, soul, and body. The Holy Spirit of God then works in us to totally separate us unto God in our spirit, through our souls, using our bodies as instruments of God's work. Adam and Eve, before their fall, were spiritual beings. This fact was established by the reality that they did not know they were naked. Christians, after their salvation, once again become spiritual beings growing in their awareness of their spirituality and becoming less and less controlled by their flesh. This is why Christian rehabilitation works so well. It deals with the real source of our problem, which is being dead spiritually. Coming alive in our spirit by God's Holy Spirit is the only way to truly bring the flesh under control.

CHAPTER 10
WHICH FALL?

The title of this chapter is stated as a somewhat sardonic question. Sardonic because evolution based worldviews confuse the circumstances of the fall to the point that the reality of a fall becomes clouded. Case in point, if man evolved from the animals and is not a higher creation, he has nowhere from which to fall. If death has always been a part of the progression of life, then, when did sin become defined, or at least, how can we really know how bad sin is? If sin did not bring death into the world, then the payment for sin was not necessary. The redemptive work of Jesus to pay the penalty for our sins becomes meaningless. If there was no fall, then how do we find authority to explain the reality of good and evil? If there was no fall and God created through evolution processes that include death, pain, suffering, and competition between species, then how can we believe a god like that would create a new heaven and earth to deliver us from these creative processes of evolution? We would have to believe it would be more of the same.

Evolution does not provide a historical or philosophical reason or need for a fall. Evolution worldviews understand basically that the way things are now are the way they have always been. The Genesis record of history establishes the fact that all was different before the fall. Creationism is the only worldview that provides an explanation for a fall and thus provides a reason for God to reverse the fall by providing a Savior.

Creationism then, provides the basis for the hope that God will restore things back to a life that was much better before the fall. On the other hand, evolution, on a philosophical level, can only promise the continuation of the way things have always been. There is no good to go back to in evolution theory. The evolution process is all there is.

Albert Einstein was a Jew who grew up going to Christian schools in Germany. He became an atheist as a young man after being heavily influenced by naturalistic science and being disillusioned by the ravages of war and anti-Semitism. After he was famous, he became a believer in a god because his theory of general relativity required that the universe had to be expanding from an original point. This meant the universe had to have had a beginning and so logic required Einstein to accept that there had to have been a beginner. Sadly, he never came to believe in the God of the Bible because evolutionistic theologians could not answer one most important question he had. The question was, "if God is a good God, then why is there so much pain and suffering in the world?" Only a literal creationist can answer that question from a strict historical understanding of a good creation that is fallen. A good creation that included choice and accountability that led to a fall that separated us from a good God. That literal explanation always leads to the reasonable solution that God, in His goodness, has provided a way out of this pain and suffering, resulting from the fall, through His Son, Jesus Christ.

An evolutionistic theologian can point to Jesus Christ as our redeemer from the New Testament teachings about Jesus. On the other hand, he does not have a historical base of reality to back up his arguments for the need of a redeemer that would cause God to provide a redeemer.

It is amazing to me how so many intelligent thinkers never think their presuppositions through to their logical conclusions. Without a fall, a redeemer is illogical and unnecessary. Einstein did think the assumptions of his worldview through and remained a Spinosa deist until his death. Spinosa taught that a god created the universe but left it to go and do other things without any further concern of what took place on earth. Einstein remained committed to an old earth, evolution worldview rejecting the Biblical God. His evolutionistic worldview would not allow him to see how an evolutionistic god could be benevolent.

THE SERPENT

Chapter 3 begins with an introduction of the serpent. Here again we must keep in mind that God's word is its own worldview. The serpent once again is another example used by skeptics to prove this account to be a mythological story rather than an event in history. Their reasoning is that snakes do not talk in real life, that is, as we know it now. But, the Genesis account makes it clear that things have not always been as they are now. That account tells us that there was a creation that was perfect and then a fall took place that changed everything. After that, there was a flood and everything changed again. We also know from the Biblical worldview that an angel talked through a donkey to a prophet and demons sometimes spoke through humans. Here we see a serpent used as an instrument by Satan to speak through.

The narrative tells us that the serpent was, " ...more crafty than any beast of the field which the Lord God had made." We are told three things about this animal by this verse.

First, he was "more crafty" then the rest of the animals. We can infer from this description that the serpent was the kind of animal , because of its clever ways, that the humans would be drawn to because it was fun to be around, more so than the rest of the animals. We are intrigued by monkeys because they express a degree of intelligence. The same is true of dogs and cats. Henry Morris tells us in the notes in his *Defender's Study Bible* that the Hebrew word for crafty or subtle, in the *King James Version* of the Bible, is *nachash* and possibly originally meant "a shining, upright creature."[33] Shining may infer being brilliant for an animal or physically appealing. The serpent was Satan's perfect choice because it was the one animal to which Adam and Eve were drawn and communicated with most, in all probability. Second, the serpent was a beast of the field. He may have walked upright with legs that allowed it to be face to face with man before being cursed to crawl on the ground. This would mean that it was not a small animal but possibly large in size. We know that Satan is sometimes referred to as the serpent of old or even as a dragon, as in Revelation 12. I was intrigued by the movie, *Dragon Heart*, which was a story about a talking dragon trying to get his heart back. It reminded me of this account of a possible good dragon turned bad. As we have already seen, myths usually had their beginnings in real events. Third, this animal was a part of the original creation and included in the pronouncement that the creation was "very good." The animal was not evil in itself. It became a symbol of evil after being used by the "evil one" for his own wicked purposes.

The first verse of chapter 3 goes on to say, "And he said to the woman." We do not know specifically how the serpent spoke to Eve. We know that before the fall Adam and Eve had dominion over the animals in a way that is not the same today.

The serpent may have spoken audibly; he may have spoken telepathically, or Satan may have spoken through the animal as a cover-up. Remember, the serpent was not evil at this time but Satan was already fallen, making it clear that the serpent was used by the devil. We have to keep in mind that both Eve and Satan were spiritual beings at the time. Adam and Eve were not yet fallen and so they could have possibly communicated on a spirit level. We also know that Satan was the source of evil and power behind the king of Tyre in Ezekiel 28 and the king of Babylon in Isaiah 14. He will have the same relationship with the anti-Christ introduced in Revelation 13:4. Satan also took control of Judas after he was convinced to betray Jesus. It is interesting to note that the only places we see Satan confronting anyone directly face to face is in Job 1 and 2 when he confronts God about Job and in the gospels when he tempts Jesus in the wilderness. Remember, Jesus was alive in spirit when tempted by Satan just as Adam and Eve were spiritual beings before the fall, when they were tempted by Satan. My point is that Eve and the devil's conversation could have been on the spirit level.

The temptation of Jesus was very similar to the temptation of Adam and Eve. One difference was that Satan came to Jesus undisguised because a disguise would not work with Jesus. Satan would have to have come in the dulled beauty of his former splendor to try to get Jesus to join with him and the fallen angels in their rebellion against God. Satan's attempt with Jesus was far more direct than it was with Adam and Eve as we will see. The goal was still the same. Get Jesus to rebel against His Father and abdicate His position to Satan just as Adam and Eve were deceived to do. This is all consistent within the Bible's own worldview.

What we do know is that when the serpent spoke, it did not come across to Eve as being abnormal. It seems that this conversation may have been going on in a normal everyday way and then turned diabolical before Eve was able to discern what was going on. She certainly was caught off guard. It reminds us of Paul's warning to be alert to the schemes of the devil. Peter was one of Jesus' closest friends and yet Jesus had to say to him one time, "get behind me Satan..." in Matthew 16:23. Peter was trying to tell Jesus that He would not have to go to the cross; Jesus recognized immediately that it was the devil talking through Peter's man centered motives. We all encounter times when we are tempted to do things we know are contrary to God's word. That is when we need to shout like Jesus did, "get behind me Satan." Don't we all wish Adam and Eve had done the same thing.

SATAN

Although Satan is not mentioned here by name, the rest of the Bible understands that the real culprit who brought about the fall was Satan, without a doubt. Revelation 12:9 is probably the most clear statement of this fact, "And the great dragon was thrown down (from heaven), the serpent of old (the fall) who is called the devil and Satan (adversary), who deceives the whole world..." Jesus spoke of Satan as he rebuked the Pharisees for their hypocrisies by saying, "You are of your father the devil, and you want to do the desires of your father. He was a murderer from the beginning, and does not stand in the truth, because there is no truth in him. Whenever he speaks a lie, he speaks from his own nature; for he is a liar, and the father of lies." The "beginning" would refer to the temptation of Adam and Eve which led to their death making him a "murderer."

His whole conversation with Eve was based on lies that actually inferred that God was a liar. These being the first lies in history make Satan the father of all lies because lies began with him.

Ezekiel 28:12-17 gives the most extensive description of Satan. This passage starts out as a prophecy against the king of Tyre but the context makes it clear that the real source of the king's evil is Satan himself. I will summarize some of the important things said about Satan in this passage without being exhaustive. Originally, Satan was the most beautiful angel of all. In other passages he is referred to as, "an angel of light" (2 Corinthians 11:14). He was the head of all the angels or at least the cherubim who guarded the throne of God and reflected His glory. He was in Eden (as we already know from Genesis 3) as well as in heaven. He apparently was free to go back and forth between the two as we see in Job 1 and 2 where he presented himself to God after combing through the earth.. Satan was, "blameless in all his ways from the day he was created..." suggesting his creation took place during one of the early days of the six day creation along with the hosts of heaven. All of his beauty, position, power, and intelligence caused Satan to become full of pride, causing him to want to become equal to God. Rather than worshiping God and reflecting God's glory, Satan determined to be worshiped himself. He over looked the fact that he was a finite created being. A nothing compared to infinite God.

A similar passage in Isaiah 14: 13-14, where the king of Babylon's source of evil is described to be Satan, expresses his evil heart. This passage says, "But you said in your heart, I will ascend to heaven; I will raise my throne above the stars of God, and I will sit on the mount of assembly in the recesses of the north. I will ascend above the heights of the clouds; I will make myself like the Most High." It is possible that Satan got the idea that maybe God had evolved from a lesser god to the God he had become and so it was possible for him to become a god as well. It seems that for someone like Satan, who lived and had his being in the very presence of God, familiarity bred contempt. We are told in other scriptures that a third of the angels fell along with Satan. They too had their own selfish desire for position, power, and glory. As we will see, this was not true of Adam and Eve. They were content with their relationship with God until Satan deceived them. Nonetheless, there are still those of mankind today who determine to rule their own lives by being their own god, just as Satan did many years before. The fact that Satan corrupted himself before the very face of God as opposed to man being deceived, may be the reason why redemption was not made available to Satan as it was to man.

THE TEMPTATION

An important question we can ask as we consider the temptation of Adam and Eve by Satan is how much time elapsed between the time God created Adam and Eve and the time they were deceived by the devil and forced out of the garden? When you read the text as it is written, you get the idea that it was the next day or at least almost immediately. That does not necessarily have to be the case. It seems that there needed to be some time to transpire in order for Adam and Eve to become vulnerable to temptation and for Satan to begin his rebellion against God. Genesis 5:3 tells us that Adam's son, Seth, was born when Adam was 130 years old.

Genesis 4:25 tells us Seth was born after Cain murdered Abel. We know Cain and Able had to be grown men from what we read about them in Genesis 4. If Adam and Eve lived in the garden 100 years, Cain could have been as old as 29 when he killed Abel, giving Seth another year to be born after Abel's death. It did not have to be 100 years, but there was a time frame established that made 100 years living in the garden before the fall possible.

Another important point to make is that Adam and Eve were not allowed by God to have children until after they were tested by the devil's temptation. We know that God is totally sovereign over the birth of a child. We see this clearly in the births of Isaac, Esua and Jacob, Joseph, Samson, Samuel, and John the Baptist. All their mothers were barren before God allowed them to conceive and give birth. It seems that God intended for Adam's descendants to inherit the positive or negative nature of Adam from the beginning based on the choice he would make while being tempted. That would make it clear that because all were in Adam when Adam sinned then all sinned. All of Adam's descendants would be sinners if Adam chose to sin or all of his descendants would be righteous if he chose to obey God. This reality would then carry over to all those who would be born again in Christ. When He died and rose again they would be declared righteous in the righteousness of the second Adam. Paul makes this clear in Romans 5:17 and I Corinthians 15:48 that all those in Christ receive His spiritual genes. The principle of those being in the loins of a forefather having their forefather's actions applied to them is found in Hebrews 7:9-10. In these verses, Melchizedek is shown to be of a higher priesthood than that of Levi. The argument is that Levi was in the loins of Abraham when Abraham gave an offering to Melchizedek.

The implication is that Levi gave an offering to him as well. In the same way, all who place their trust in Jesus become in Christ and all that applies to Christ applies to those who are in Him. In Adam, we all have sinned. In Christ, we are all declared righteous.

Satan, through his scheming ways, began his temptation by putting a question in Eve's mind; a question she would have never thought of herself. Up to this moment, neither Adam nor Eve had had reason to question God about anything. The only questions they had concerned their learning about how the creation worked. "Indeed, has God said," Satan asked? He knew, by his crafty fallen way of thinking, that questioning God's word is the quickest way to plant seeds of distrust that lead later to bad choices. He certainly was setting Eve up. It has been a major part of his tactics from the beginning. It is important to see that at first glance the question, from Eve's point of view, would not have been dangerous in that it called for clarification. It did not agree with what God had said. Eve, being highly intelligent, but totally naive about evil, never suspected that Satan's twisting of the truth was only a ploy to move the conversation to his diabolical deception.

Listen to how Satan words his question, "Indeed, has God said, 'You shall not eat from any trees of the garden?'" Notice how extreme it is in being totally wrong in how Satan quotes God. God only denied them use of one tree but Satan includes all the trees. It seems he is baiting Eve to quote God as saying they would "die" if they ate from the one forbidden tree. In fact she missed quoted God as saying, "You shall not eat from it or touch it, lest you die." God did not say, "or touch it."

That was not a bad idea to not even touch the tree, but that was not what God said. Nevertheless, God did say, "You shall surely die."

"Die," that was the word Satan was looking for. First, Satan causes Eve to question God's word; now he is going to cause her to question God's integrity. He responds to Eve immediately with, "You surely shall not die." Here we see Satan quote God word for word but adds the one little word, "not," that totally reversed what God actually said. Without saying it directly, Satan was calling God a liar. Eve was then confronted with the test of her life. She had to decide if she was going to trust God by believing what He said, or if she was going to listen to the serpent and believe what he had just said. Tests come in life at times when we least expect them. I am sure that, these many centuries since, Eve now knows Satan is a liar, but that was not the case back then. Sadly, Eve chose to listen to his first lie, which led to an even greater lie. Keep in mind that Eve was not accustomed to being lied to. She was not aware of what the devil was up to, she did know what God had said in contrast to Satan's lie. It does not take long to get caught in Satan's traps if we are not alert to his schemes. When I watch a documentary on snakes and they show a snake slithering quietly and slowly up to its prey unnoticed and then, suddenly, it strikes without mercy, I think of this passage. Eve was the first serpent's unsuspecting prey. All the pain, suffering and death in this world began as the result of this event. Nothing has changed. Satan is still the serpent and we still are the prey, if we do not learn to trust God's word and learn His ways. Knowing the ways of God is our greatest defense that alerts us to the schemes of the devil.

At that moment, when Satan called God a liar, Eve should have run for all our lives and Adam with her, just as Joseph did when confronted with the temptation through Potiphar's wife. Maybe this passage was where Joseph learned to run when confronted with evil. But the fact of the matter is, Eve did not run. Instead, she stayed to listen to more of Satan's lies. That is why it is so important to know what God's word says. Knowing God's word is our hedge of protection against Satan's schemes and lies. When we read or hear anything that contradicts God's word, we must reject it immediately. That is how we protect ourselves from false teachers and organizations that are Satan's instruments of deceit. Just as the serpent in the garden became Satan's instrument of deceit, he still uses ploys that seem harmless at first glance. Often times, what seems so harmless in the beginning can end up being the death of us, if we do not know how to recognize a snake when we see it. Just ask someone addicted to alcohol or drugs.

Satan's strategy was working. He put a question in Eve's mind then he attacked God's integrity by calling Him a liar. Before Eve could recognize what was going on, the devil embellished his lie by using a half truth to strengthen his accusation. He goes on to say, "For God knows that the day you eat from it your eyes will be opened, and you will be like God, knowing good and evil." It is true that Adam and Eve's eyes would be opened to see good and evil, but only because they would die to good spiritually and become evil themselves. They would continue knowing good but be totally unable to produce that good any longer. They would know good but only from an evil and fallen position.

The sad thing about this temptation is Satan was tempting them to become something they already were. He said, "...you will be like God..." The fact of the matter was, they could never become more like God than they already were. The big difference was God had protected them from having to know evil. God only wanted them to know good, not evil. The great paradox here was Adam and Eve were tempted to become what they already were, and that, by a lesser being than themselves. They were made perfect in the very image of God. Satan was a fallen angel who was created to serve God and man. Now they were to become subject to this egotistical tyrant because they had fallen under his control rather than God's. By listening to Satan, they abdicated their place of power to him and allowed the devil to become the god of this world. For a season, man became subject to a fallen angel. That angel was a lesser being who was intended by God to be our servant. The wonderful thing is that the Son of God was willing to become our servant to restore us to God's original intended position of ruling with Him.

After listening to the devil's lies, Eve began to gaze on the fruit of the tree of knowledge of good and evil. Because of Satan's false information, she saw the fruit in a way she had never seen it before. She must have reasoned, "Well it is good for food and it truly is a delight to the eye, how bad could it really be? After all, God did create it, and I certainly want to be as wise as God." Satan will always help us reason our way into sin. Those reasoning processes are familiar to us all, such as, "after all, it does feel good; after all, we do love each other; after all, we are not hurting anyone; after all, who is going to know; after all, it is my life and I can do what I want to with it."

Before we know it, we have taken a bite and it is too late; sin has separated us from God. Chapter 3:6 says this, "When the woman saw that the tree was good for food, and that it was a delight to the eyes, and that the tree was desirable to make one wise (according to Satan), she took from its fruit and ate; and she gave also to her husband with her, and he ate." Now the damage was done. Sin was complete. Satan had succeeded with his deception moving one step closer in his plan to supplant God (Isaiah 14:14).

Notice the progress of the temptation that finally won Adam and Eve over. First, we see that she lingered to look and let what Satan had said sink in. The fruit then, as she looked at it from the perspective of a lie, appealed to her appetites. She saw it was good for food. Second, she was drawn to its appeal to pleasure. It was a delight to the eyes. Third, it appealed to her pride. It was desirable to make one as wise as God. As a spiritual being, she warred within herself not to listen to Satan just as God's Spirit wars in us when we are tempted. In her mind though, she was convinced, like Satan, that it was a desirable thing to be her own god or at least equal to God. Her mind reasoned the fruit was good for wisdom, pleasure, and food. Her body certainly agreed with what her mind was thinking. Her emotions were led to believe by the devil's lies that God was holding out on her.

By her will, she chose to direct her physical body to eat, something her body was already desiring to do through its appetites and desire for pleasure. Tricked by false reasoning, influenced by pride, pleasure, and appetites, Eve chose to disobey God. Her sin, as a result of her bad choices, ruined everything.

Human pride, worldly pleasures, and uncontrolled appetites have undermined freedom, trust, and love ever since that first sin committed through our fleshly desires. But, we are still responsible and accountable for the bad choices we make. It was this tri-level temptation John refers to in I John 2:16 when he wrote, "For all that is in the world, the lust of the flesh (appetites) and the lust of the eyes (pleasure), and the pride of life (pride to be one's own god), is not from the Father, but it is from the world." The challenge of resisting these temptations remains today.

The big question is where was Adam while this deception was going on? When discussing the temptation, I have often included Adam as being tempted along with Eve even though the account gives the impression only Eve was there speaking to the serpent. I have done this for several reasons. First, the statement, "...and she gave to her husband with her, and he ate..." suggest that Adam was not far away and possibly there listening to the conversation. The fact is, Adam did not speak up for God, but chose to remain silent and even was convinced to eat the fruit himself. Second, we will see that the eyes of both Adam and Eve were not opened to their nakedness until after Adam ate. This tells us that because of Adam and Eve's oneness pronounced in chapter 2, their sin was not complete until both of them ate. Third, it was Adam who was given the command not to eat of the forbidden tree. The command came before Adam named the animals and Eve was formed from his side. Apparently, Adam was ultimately the one held responsible by God for the command and he gave it to Eve possibly adding the, "... or touch it..." that was not a part of God's original command. When it came down to it, the buck stopped with Adam. It is still true today.

The man in the family is responsible to God for the spiritual oversight of his family. Adam failed because he did not stand firm on what he knew God had commanded. We all do the same when we fail to obey God's commands.

QUESTIONS

We have some questions to answer in light of the temptation. The first question is, why did Adam choose to eat rather than say "no" to Satan's temptation? We are told Eve chose to eat the fruit before she gave it to Adam. First, she ate and did not die. Actually, nothing happened to her after eating the fruit and before giving it to Adam to eat. This would have affirmed that Satan was right when he said to Eve that they would not die. This would have given him more confidence to believe the devil's lie, not knowing that sin's consequences often are a delayed reaction. Second, and even more important, the fact that Eve had already eaten put Adam in a dilemma. He was very much attached to Eve but she had disobeyed God's command. Adam had to choose between Eve and God. He chose to take Eve's side, and in doing so, later received the rebuke of God in verse 17 where God said, "Because you have listened to the voice of your wife... ("and not Me" is inferred)." It may have been that Adam, because of his love for Eve, may have thought that he could save Eve by joining her in sin. How often do Christians make the mistake of thinking they can marry an unbeliever in order to help them get saved. A new Christian may continue to go to places with old friends thinking they can help their friends come to Jesus but get dragged back into sin instead. The scripture says to, "come out from among them," for good reason.

I believe the main reason why Adam chose to join Eve in her sin was because he fell prey to the same temptations Eve did, the lust of the flesh, the lust of the eyes, and the pride of life. He was just as hooked as Eve was. He wanted to be equal to God.

The second question we need to answer is, what if Adam had chosen not to sin, what would have become of Eve? I believe we have the answer to that question in the illustration Paul used to compare how a man is suppose to love his wife like Christ loved the church in Ephesians 5:25-27. Christ saved the church (all believing sinners) as his bride by obeying God and sacrificing himself on her behalf. Adam may have tried to save his wife by joining her in her sin, when, if he had obeyed God and chosen not to eat, he possibly could have saved both himself and Eve. If you look in the next verse after Adam's eating of the fruit, God's word says, "Then the eyes of both of them were opened..." The "then" tells us that their eyes were not opened to their fallen state until after Adam ate. We need to remember God had pronounced them "one flesh." If Adam had not eaten the fruit, their sin would not have been complete, by his obedience. Adam could have saved Eve and himself by obeying God. Instead, because he listened to Eve (and Satan's lies), he brought God's judgment on himself and Eve together. Thank God the second Adam, Jesus Christ, chose to obey His father and, by so doing, saved His bride, the church, made up of all those who believe in Him as the Son of God.

A third question that comes to mind is, What did Satan hope to gain by causing Adam and Eve to fall? We have already seen that Satan's major objective was to become equal to God and place his throne above the stars (angels) of God.

(It is interesting to note that Satan desired to be equal to God not greater than God. Why? Because, even Satan had enough sense to know no one could become greater than God because God's greatness is infinite.

You cannot get any greater than that. Be that as it may, we will get back to our question.) Satan had already convinced a third of the angels to join him in his rebellion, but angels did not hold the same place before God as man did. It seems that the devil may have reasoned that if he could get man to fall under his authority, he could use man as leverage against God.

Think of it like this. Man was the ultimate object of God's creation. He was God's prize creation. Satan, in his twisted way of reasoning, may have thought he could get God to violate His own justice and holiness by putting God in a compromising position of having to choose to destroy man or let him off the hook. Knowing the great love of God, Satan deluded himself into thinking God would compromise Himself in order to save man. But there was something Satan did not know about God's plan for man. Before Satan was even created, God had already planned to crucify His own Son on behalf of man after man sinned. As God's word says in Revelation 13:8, this was God's plan, "from the foundation of the world." When Jesus shed His blood for man's sin, Satan's plan was destroyed. We see in scripture that Satan (the accuser) continued to accuse the saints of God until Jesus shed His blood for our sins. After that sacrifice, Satan, in Revelation 12: 9-12, was kicked out of heaven. Why? He could no longer bring any accusations against God's children of faith in Christ and so there was no longer a legal justification for him to approach God's throne.

After being kicked out of heaven, Satan knew he was defeated and his days were numbered. Since that time, he has had nothing better to do than to work to bring as many to hell with him as he can deceive. That will be the devil's only consolation in eternal hell. There will be other fools there with him who thought they could be equal to God.

Another question I want to try to answer is highly philosophical and yet important to consider. The question is what is the significance of Satan anyway? The best answer I have come up with after reading, studying, and meditating is God, being good and yet allowing choice, made it possible for created existence to become evil. If evil was to be a consequence, God would have to deal with it and finally do away with it once and for all. That would make it possible for Him to have a creation that was void of evil without the possibility of it ever tainting His prefect creation again.

The ultimate purpose of this first creation is to personify evil and do away with it once and for all in a final eternal judgment. Satan and those who choose to join him in his rebellion, whether angelic or human, out of choice, have become the personification of evil that God will finally separate from His good creation forever. There will not be an opportunity for evil to infiltrate God's new heaven and earth ever again because all those in that new existence will have already made their choice for good. The choice for evil (against God's good) will no longer be necessary. Those relegated to hell will be there because they chose to love darkness rather than light or good. They chose for evil and so will never again have the choice for good.

The sad thing is, they will long for that opportunity to choose good for all eternity but that opportunity will never come again. Through all these rebellious ones, evil will have been personified and dealt with once and for all never to be a threat again.

EVIL

This discussion of evil brings up our final question. What is evil? If God is good then evil could have no part in His existence and definitely something He would not create. Pantheistic religions of the world try to say that good and evil are the same thing. They are opposite sides of the same coin. What is good to some may be evil to others. The reality is that our own rationality tells us that good and evil exist and they are opposites. So, where did evil come from? First, God did not create evil; evil is a consequence of choice. Choice made evil possible but evil was not created by God. Second, Adam and Eve were created totally good as was Satan in the beginning. We will see in the next section that when Adam and Eve's eyes were opened and they knew they were naked something radical changed about their existence. They did not die physically, although they began to die or age from that point on. But, God did say that in that day they ate of the forbidden tree they would surely die. The reasonable thing to assume then is that they began to die physically but they did die spiritually immediately that day after they ate the forbidden fruit. The Bible makes it clear in many ways that, because of the fall, all humans are dead spiritually in their trespasses and sin. Jesus, in John 3, makes it clear that salvation is the act of being born again or once again being brought back to life spiritually. Paul tells us in I- Corinthians 2:14

that a natural man or one who is not alive spiritually cannot understand spiritual thoughts.

What is my point? When Adam and Eve sinned, God removed His presence from them. They died in their spirit because the Holy Spirit left them. Left to themselves, they became evil, giving us the foundation for understanding what evil is. Evil is the absence of the presence of God's good. Only God is good. When Adam and Eve chose to be their own god, God removed the good of His presence from them and, left to themselves, they became evil.

Not only did Adam and Eve become evil, but the whole creation itself became evil for the creation was created for the habitat of man. When man was good, his habitat was good. When man became evil, his habitat had to become evil as well. That is why Paul wrote in Romans 8 that man's habitat continues to groan to be delivered from that evil, or more clearly, to experience the return of God's good. The earth itself is groaning to be delivered from the burden of the consequences of man's sin.

I want to give you an illustration of this explanation of evil that comes from the life of Albert Einstein when he was still a young student. One of his university professors asked the class if God created everything that exists? One Bible believing student confidently replied that He did. The professor went on to make the point that if God created everything, then God had to have created evil because everyone knows evil exists. The professor then said, "If our works define who we are, then God is evil." The professor was proud that he had proved once again, to himself at least, that Biblical faith was a myth because of the existence of evil.

Young Einstein then asked permission to ask the professor a question. "Sir, does cold exist?" The professor answered in the affirmative a little taken aback by a question that had such an obvious answer. Everyone knows cold exists. Einstein replied, "Sir, cold does not exist. Cold is the absence of heat. Heat can be measured but cold cannot be measured. Once all heat is gone, cold is all that is left." Einstein went on to make the similar point about darkness. Darkness is not measurable. Darkness is the absence of light. Light can be measured but darkness can only be measured by the amount of light present. After making this point, Einstein then showed that we can only measure evil on the basis of what good is.

"Evil is the absence of God," Einstein said, giving his professor much to think about.

The same thing can be said of life and death. Death is the absence of life. You cannot measure death, you can only measure how much life remains in a body. Once life is gone, a dead body is dead. Dead is the absence of life. Nothing can be more dead than something else that is dead. In the same way, evil is the absence of the presence of God's good. Evil is not created, it is a result of the removal of good just as cold is made possible when heat is removed, and darkness becomes possible when there is no light. In the new heaven and earth, there will be no cold because there will only be God's energy; there will be no darkness because God's light will perfectly fill every corner with light; there will be no evil because God's good will fill all existence there; and there will be no death because God is the living God who gives life and that more abundantly. Sadly, those in Hell will only experience the presence of God's wrath.

C.S. Lewis was an atheist who converted to Christianity after discovering that the arguments against Christianity and for atheism could not be supported philosophically.

As an atheist he would use the argument of the existence of evil to disprove the existence of God. He reasoned that evil makes a good God impossible. As he reasoned his own argument out, he discovered that the existence of evil actually requires a good God to exist. If there is no good God Who is the absolute reality of good, then good has no philosophical reason for its existence. If there is no absolute good, then there can be no evil because there is no moral basis for either concept. Good and evil becomes relative to what each individual or society decides what they want them to be. Postmodernism would say neither good nor evil exists.

They are non-entities. Lewis was convinced there are certain things that are good and others that are evil on the basis of natural law and so he came to accept the existence of God because good and evil require an absolute good for either to exist.

There is no higher authority that exists in atheism to define good and evil for a society. Their definition is left to those who have the power. Thus in Germany after World War II the Nazis on trial for war crimes could justify the Holocaust on the basis of their own culture's definition of good and evil. For them, it was good to protect the purity of their race by doing away with those lesser races who threatened to weaken their own. Genocide was justified by their godless evolutionistic worldview that established survival of the fittest as their moral responsibility. Without the existence of a good God, anything can be justified by society no matter how horrific.

Where there is no moral authority there is only anarchy where only the strong survive in an immoral chaos. As Ronald Reagan once said, "A nation that is no longer under God is a nation gone under."

VERSE 7

Genesis 3:7 gives some transitional information that takes us to that moment of accountability that always comes after sin has had its way. Verse 7 says, "Then the eyes of both of them were opened, and they knew they were naked; and they sewed fig leaves together and made themselves loin coverings." We have already discussed how this verse describes the death of Adam and Eve's spirits by their becoming aware of their nakedness. Their immediate death was spiritual, which would lead ultimately to their physical death.

All of a sudden, what was once taken for granted as good (nakedness) now became seen as evil. Their awareness of their physical bodies exposed their fallen state and they became ashamed. This verse also gives evidence that this account was written after the fact of the fall to explain why Adam's readers experienced shame and why being lewd was wrong before a holy creator God.

This verse explains why humans wear clothes. An explanation for wearing clothes is not found anywhere else in human cultural development. If we evolved from animals, why do we feel shame for our nakedness? Where did modesty come from? Some would say clothing was developed as a practical tool to protect us from the elements, but that does not explain the reason for shame and modesty. It seems this historical record of the fall provides the original explanation. Even though we find some primitive cultures dead to shame, showing themselves to have developed an inner resistance to the consciousness of their own fallen nature, it is interesting to note that when primitive cultures become Christians, they become modest and start wearing clothes.

This was true of the Gaderean demoniac who was fully clothed and in his right mind after encountering Jesus. The truly cultured person has not lost the understanding of modesty because his or her conscience remembers the reason for that modesty.

We are fallen and the shame of our nakedness is a continual reminder of our fallen state. There are primitive cultures today in remote areas of the world who dress very scantily with nothing more than a pubic covering. It is interesting that the oldest cultures used clothing, including primitive cultures that lived in caves for a time after the flood. Those cultures today that dress very scantily would be an expression of man's loss of original sensitivity to our fallen state. They also show the wearing of clothes did not evolve out of necessity for protection. Thus, clothing is more a result of shame rather than necessity for protection.

The fact that Adam and Eve felt a need to cover themselves reveals something very important about them. They had developed a sense of shame. Feeling shame was a positive for Adam and Eve in their fallen state. Why? Because shame shows an awareness of sin and that shame can lead to repentance. It is a dangerous thing when a person or even a nation losses their sense of shame. Jeremiah makes this point in his prophecy to the people of Judah when he wrote in Jeremiah 8:12, "Were they ashamed because of the abomination they had done? They certainly were not ashamed, and they did not know how to blush; therefore they shall fall among those who fall; at their time of punishment they shall be brought down, declares the Lord."

This is an ominous warning to once vibrant Christian nations who have allowed pornography, fornication, adultery, homosexuality, abortion, greed, and many other abominations to become common practices in their societies. When a person or nation no longer knows how to blush, they no longer know how to be ashamed. Without shame, they will not repent; leaving God no alternative but to bring His judgment. The history of the rise and fall of nations from the beginning confirms this fact.

DEPRAVITY

The words "knew they were naked" are words expressing a new kind of knowledge. It seems the concept of the knowledge of good and evil had taken over in the fallen man and woman's experience. What was once good from then on became known as evil. They knew good and evil but after sin came they knew the difference only from an evil perspective. Man, having become evil, was no longer capable of producing any good that was acceptable to a totally good God.

The only good acceptable to God is the good that only God Himself can produce. Any good that man tries to produce will always be tainted by his fallen nature. As Jesus said, "a bad tree can only produce bad fruit." This runs head on into the secular humanist understanding that man is basically good and left to himself will do what is good. No, man is fallen and has to have laws, police, jails, and the fear of punishment to keep his depravity under control. A child does not have to be trained to be bad. It always has to be trained to do good. As the scriptures say, "There is none righteous no not one" and "Our righteousness is as filthy rags before God." The concept of the total depravity of man is very important.

The fact that man is totally fallen and incapable of producing any good that is acceptable to God is the foundation for understanding why we all have to be saved by grace as God's free gift. God had to save us through the righteousness of Jesus because we cannot produce any righteousness from within ourselves. Man cannot save himself by his own works because those works are smudged with sin. Salvation has to be a free gift because we have no goodness to purchase it for ourselves. God has declared Jesus Christ as righteous by His resurrection.

Those who accept Jesus as their Savior are declaring their own lack of righteousness before God and claiming the righteousness of Jesus as their only hope. A person who is in Christ Jesus is declared clean by God. No other religion in the world teaches our need for God to save us by His finished work. They are all based on man working to develop his own righteousness by "good deeds." That is impossible because man is no longer good according to God's standards. Biblical salvation is God reaching down and pulling us up to Himself by His goodness.

World religions are self helps on how man can pull himself up to God establishing his own goodness. The Biblical worldview teaches that man trying to make himself acceptable to God on the basis of his own goodness is an exercise in futility. Why? Because man is not good. Biblical salvation is, "...not as a result of works, that no man should boast." (Ephesians 2:9)

The next part of verse 7 is pathetic. It is a perfect picture of how man tries to find a way to cover up his guilt in his own eyes and before God. The passage tells us that Adam and Eve tried to cover up their nakedness with fig leaves taken from a tree in the garden.

This is an illustration of how our works are such feeble attempts at trying to reverse the irreversible. Man tries to cover up the result of his problem, his nakedness, rather than realize the root of his problem, his spiritual deadness. A good example is man dealing with his addictions. He comes up with all kinds of "step" programs to bring his appetites under his control failing to realize that, until he is brought back to life spiritually, he is fighting a losing battle with himself and his own fallen nature. He is his own worst enemy. The problem for Adam was not his nakedness; the problem was he was totally dead as a spiritual being. Like Adam, so often we try to fix those things only God can fix for us. Only God can restore us to spiritual life in Him.

THE CONFRONTATION

The damage was already done. There was no turning back. All that was left was the foreboding thought that sooner or later they were going to have to face God. That is the thing about God; He never goes away. You can deny He exists. You can say He does not care and has nothing to do with His creation anymore.

You can convince yourself that God is love and cares nothing about our sin. But something within us tells us God is coming and we will have to face Him in time. For years it was the fear of facing a holy God that caused Martin Luther to torture himself with every form of self inflicted penance in hopes of finding cleansing before his ultimate confrontation with the Creator and Judge of the universe. Thank God, he learned and helped the church to relearn about salvation by grace through faith. Adam and Eve had not yet learned this truth.

The next words in Genesis 3:8 remind me of the childhood situation when children know they have done something wrong while their parents were away. Mom and Dad were gone for an evening out but all of a sudden they drive up in the driveway sooner than the kids expected. There is no time to fix things and so the only thing to do is to try to cover up the evidence and hide. Being unsuccessful in their cover-up, their guilt is obvious. Adam and Eve find themselves in a similar situation in verse 8. The verse says, "And they heard the sound of the Lord God walking in the garden in the cool (or calm breeze) of the day..." This passage suggests to us that God took on a physical form of some kind and visited with Adam and Eve in the garden often. This anthropomorphic transition was probably similar to the angels that came to visit Abraham before going to destroy Sodom and Gomorrah. They looked like men but Abraham knew they were angels. None the less, this was God making a sound as He walked in the garden. The truth was obvious, there was no hiding from God after Adam and Eve had sinned.

We have to ask ourselves an important question at this moment in the account. Did God know Adam and Eve had already sinned? Of course, God is all knowing.

He knows the end from the beginning, and so there is no doubt God already knew what had taken place. What does that tell us about God?

God seeks us out; even when He knows we have denied Him; even when He knows we have rebelled against Him; even when He knows we have rejected His authority; and even when He knows we would never seek for Him. Left to themselves, Adam and Eve would never seek for God in their fallen state.

They did not seek Him, they hid from Him. It is important for us to realize that the only reason why we know anything about God is because He reveals Himself to us, not because we seek Him out. An important principle here is that we love Him because He first loved us. In other words, it was God who kept the lines of communications opened, not man. So often, God intervenes in our lives when we are at our lowest point. The last thing we want to do is deal with God, and yet, God knows that dealing with Him is what we need most, and so, He comes looking for us.

Another question that is often asked at this point is why did not God stop them before they ate? I believe the major answer to this question is God created us to be free moral agents. If He gave us the freedom to choose, then even God has to let us choose. Even if our choice is going to be the wrong one; even if it may cost millions of lives; even if it may cost the sacrifice of the Son of God; God must let us choose. We see this same scenario all through history. God often allows bad things to happen as a result of bad choices and the evil intent of sinful hearts. Often times, the innocent become the victims of those bad choices. In a sense, we could say that we are all victims of Adam and Eve's bad choice. Many people over the centuries have died needlessly at the hands of wicked men who could never satisfy their lust for power and their need to dominate others.

All this would seem totally unjust if there was not a promise from God to hold all mankind accountable for their every word and deed. There is solace in knowing that one day God will right every wrong. God's word promises us in Romans 12:19, "... Vengeance is mine, I will repay, says the Lord."

God walking in the garden solidifies the fact the He is the God of the second chance. The great message of all the prophets was "repent." When John the Baptist came to preach, he preached, "repent" (Matthew 3:2). When Jesus Christ came, He preached, "repent" (Matthew 4:17). As we will see in the following verses, God is going to give Adam and Eve the opportunity to repent. Yes, God knew they had sinned, but instead of wiping them off the face of the earth, He came looking for them. Our God, Who created us, would never leave us alone, especially in our sinful state. He may have removed His goodness but He has never left us alone. His desire is always to redeem us; to seek and to save us when lost. This verse denies the deist concept of a god who created the universe and then left it to go on and do other things no longer interested in having any part in what happens on earth. God is still walking amongst us and calling out to us while we try to hide from His presence. Why does God seek us out? All He wants us to do is repent. Why does He want us to repent? So He can grant us forgiveness. Why does He want to forgive us? So He can have fellowship with us once again and bless us.

The time for confrontation has come. Genesis 3:9 says, "Then the Lord God called to the man, and said to him, 'Where are you?'" As I have already said, God is the one who calls out to us before we ever respond to God. Left to ourselves, we would never seek for God. God's call is an intriguing question, "Where are you?" It is intriguing because we know that God already knew where Adam was.

The question is one we need to answer for ourselves rather than for God, when we find ourselves hiding from God. Adam needed to be asking himself, "why am I hiding from God?"

Introspection is the first step one has to take on the path toward repentance. Adam needed to realize that God was not his problem, his own disobedient sin was his problem. Adam was Adam's problem. In the story of the prodigal son, we see the broken young man in the pig pen confronted with the question, "Where am I?" or "How did I end up in this pig's pen?" He had to face the fact of his own bad choices before he determined for himself to repent to his father. The beautiful thing was that when he did go to his father and repent, he was completely forgiven and restored as his father's son. When we repent to God, He forgives us and restores us to being His child. God's call to Adam, "Where are you," is a call to Adam to evaluate his reason for hiding from God so he would repent.

Realizing he could not hide from God, Adam came out of hiding. It was time to face the music. Verse 10 goes on to say, "And he (Adam) said, 'I heard the sound of Thee in the garden, and I was afraid because I was naked; so I hid myself.'" Here we see the ultimate separation from God that sin continues to cause to this day. God is holy and we are fallen in sin. God is robed in righteousness and we are naked with sin. Freedom has turned to fear. The veil of sin has fallen between God and man. Evil hides from the presence of pure goodness. Adam, in this verse, then knew good and evil. He understood that God was totally good and he was totally evil in his own eyes. The only two things they still had in common at this point was that man was made in the image of God and God was still his Creator.

Soon, Adam would have to learn a new role God was going have to play in his life as his Redeemer. Verse 11 records the moment of reckoning. The question of confrontation is direct and to the point, "...Who told you that you are naked? Have you eaten from the tree of which I commanded you not to eat?"

There is a hint of holy anger in these questions, which suggest it is possible to, "be angry and sin not." We just need to be angry for the right reasons and be against the right things. God knew the answer to His first question. No one told them they were naked because they would not have been able to relate to the concept of nakedness before they fell. Up to this point in time, they were not naked as we understand nakedness after the fall. Their nakedness became the expression of their newly gained sinful nature. The second question, "Have you eaten..." gets to the point of Adam and Eve's disobedience. Did they disobey God's command?

God's question leads us to ask a question. What was the best answer Adam could have given? I believe the best answer would have been, "Yes, Lord God I have sinned." How do I know this? When Nathan the prophet confronted David with his sins against Uriah of murder, lying, and adultery in II Samuel 13, David's immediate response was, "I have sinned against the Lord." Though David was punished severely through his family, God still called David a man after His heart. Why? David was willing to repent.

All Adam had to do to show repentance was acknowledge his sin and put himself at the mercy of God. That is a true heart of repentance. Instead, what did he do? He passed the buck to Eve and then blamed God for giving him Eve in the first place. Notice what Adam says in response to God's questions, "...The woman whom though gavest to be with me, she gave me from the tree, and I ate." First, Adam blames Eve by saying, "...The woman..." and "...she gave me from the tree..." Second, Adam blames God by interjecting the words, "...whom Thou gavest to be with me.."

In other words, Adam was saying that it was all God's fault he had sinned because if God had not made Eve, Adam would not have sinned. God has been the object of blame for our sins ever since. The same excuses have been repeated over and over again. If God had not made me like this; if God had not given me the parents He did; if God had intervened when He was supposed to; if God had done what I wanted Him to do, I would not have done what I did; on and on the excuses go. There is never true repentance until we take responsibility for our own choices and actions and give up on making excuses.

Look at what has taken place in Adam's heart immediately after the removal of God's goodness from him. What love he had for God was gone. He was now on the defensive in his relationship with God by justifying himself to himself and blaming God. What love he may have had for Eve was gone. If he was trying to protect her before the fall, he was blaming her after the fall and hinting to God that she was a mistake. The truth of the matter was that Adam had become self-centered, selfish, caring for no one but himself. He was truly a changed man, but only in a negative way. It is no wonder that before a person can be used of God after salvation, God has to break us and cause us to die to self and come alive to the Holy Spirit's goodness in us. Before that happens in our lives, any good we try to do is just wood, hay, and stubble, as Paul calls our fleshly service to God in I Corinthians 3:12. It is a human good tainted with fallen self motives. Only God can produce genuine, love, joy, peace, patience, kindness, goodness, faithfulness, gentleness, and self control referred to as the fruit of the Spirit in Galatians 5:22-23. These character qualities of God's goodness are all that are acceptable to God and can only be produced by God living in control of our lives.

It is important that we see in this account that Eve was the first to eat of the forbidden fruit but Adam was the one God approached first. Adam was the one God originally gave the command to, not Eve, and Adam was the one God confronted first. This makes it clear that God considers the man to be the head of his house and responsible for his family's spiritual oversight. Nonetheless, Eve was also confronted with the part she played in this act of rebellion against God. She was accountable as well for her actions when God said to her in verse 13, "What is this you have done?" To her credit, Eve's response was more straight forward and less defensive then Adam's. She at least did not try to blame God, but admitted she had been deceived by the serpent. Here is what she said, "The serpent deceived me and I ate." There is a certain amount of her playing the blame game as well. There seems to be a hint of remorse in her answer, at least more than what Adam showed.

There is a lesson to be learned here about temptation. The comedian, Flip Wilson, made the phrase, "The devil made me do it," popular back in the 1970's. What we learn from Adam and Eve's experience is that the devil does not have the power to make us do anything. He did not make Adam or Eve eat the fruit. Satan could only tempt. Adam and Eve still had to make a choice to eat or not to eat. God does not violate our free moral agency and Satan is not allowed by God to violate it either. It is our freedom to choose that makes us responsible to God for our choices. When it comes down to the final analysis, we have no one to blame but ourselves for the bad choices we make. Our lives are filled with good and bad consequences as result of the kinds of choices we make. Thank God for His grace that helps us to overcome those bad consequences through Jesus Christ.

The big question we are faced with as we read this account of the confrontation of God concerning their sin is, "Did either one truly repent?" The resounding answer is no!

As I have already said, true repentance comes when we take responsibility for our own actions and desire to be forgiven and changed. In each case, Adam and Eve passed the blame for their actions to another. If they did not repent here, do we have any evidence that they may have repented later? Yes, there is evidence. On Adam's part we see that this whole account from chapter 2 to chapter 5 was Adam's eye witness account as Genesis 5:1 says, "this is the book of the generations of Adam." This is an account written by Adam after the fall explaining to his descendants why there are thorns and thistles, pain and suffering, death and dying in a world created by a good God. Adam was admitting that things are the way they are now because of his sin. In a way, he was saying, "do not blame God, blame me."

There are other indicators that Adam and Eve both repented. In verse 25 of Genesis 4, Eve names Seth and the verse tells us she saw Seth as God's replacement for Abel, who Cain had murdered. There is a hint that she saw Seth as being the seed God promised to give her to destroy the work of the serpent in Genesis 3:15. We also see that Seth's son, Enosh, began to call on the name of the Lord. Before that, we see Cain and Abel offering sacrifices to God for their sins, something they would have learned from their parents. Adam, out of his repentance, passed on the creation account and his own testimony to his children and their children, which became the foundation for their worship and hope for the restoration of God's good creation.

CHAPTER 11
WHICH CURSE?

The title to this chapter asks the question, "Which curse?" We are comparing the creation worldview to the many different worldviews rooted and grounded in evolution philosophy. There really is only one curse account, and it is found in Genesis 3. Evolution theory does not promote a curse, and yet, has become a curse itself. Here is what I mean. We have already seen that, if the evolution process of living things developing from simple to complex forms is true, than the idea of a curse as an event in history is unnecessary because death existed before sin.

That worldview is a curse, if the Biblical record of the fall is true because many people have been deceived, like Eve, into thinking there is nothing for man to be saved from, thus, making Jesus Christ irrelevant. If Christ is God's only provision for sin, but people do not believe in Him because they do not see themselves as fallen, then their worldview has cursed them by causing them to reject God's salvation. They are without hope, and that is a curse.

We saw in the previous chapter that God confronted Adam and Eve individually, but the interesting thing is, God did not confront the serpent. After Eve blamed the serpent for deceiving her, God turns directly to the serpent and begins to curse it without asking for any sign of repentance. Why is this so? The serpent was the instrument of the devil. It then, became the symbol of evil which the devil has always been the original source of.

This passage makes it clear that there is no opportunity for fallen angles, such as the devil, to find repentance. The difference between fallen angels and Adam and Eve was the angles rebelled against God in the very face of God. They were without excuse. Adam and Eve were deceived into their rebellion as a result of Satan's rebellion and deception. The implication is that they would not have rebelled if Satan had not lied to them.

Another important point to consider why the serpent was not confronted is, angels are not beings of faith. Their existence is based on sight. They live and move in the very presence of God. Their rebellion was a direct rejection of God's authority. Man does not live in the direct presence of God, and so, he has the opportunity to have a faith relationship with God. Man's rebellion was an indirect rebellion that could be salvaged by the restoration of trust out of choice through faith. Angles cannot please God like humans can because they cannot relate to God on the basis of faith.

Hebrews 11:6 makes this clear by saying, "And without faith it is impossible to please Him, for he who comes to God must believe that He is, and that He is a rewarder of those who seek Him." Man is the only being who has the capacity to relate to God by faith set by the parameters of this verse. Paul also states this requirement in Ephesians 2:8, "For by grace are you saved through faith..." Our capacity for faith gives us saving value in the sight of God because it restores us to a trust relationship with God.

THE SERPENT

Genesis 3:14 begins the pronouncements of the curses against the serpent, the woman, and the man. God starts out by addressing the serpent with, "Because you have done this..."

What is the, "this," God is referring to? It is the act of deceiving Eve and then Adam with lies that called into question the trustworthiness of God, thus leading to Adam and Eve's sin. The serpent, as Satan's instrument of deception, was the major culprit. Verse 14 deals with the curse that will continue to apply to the serpent as a symbol of evil. The serpent will crawl on its belly and eat dust as a part of its daily punishment. This verse also tells us that the serpent was to be more cursed than, "all cattle, and more than every beast of the field." Not only was the serpent cursed but the rest of the animals will be affected by the curse as well. Death and dying, pain and suffering will be their lot as well as Adam and Eve's. All of the creation will be affected by the curse of entropy leading to death for all the creation.

Genesis 3:15 then transitions in focus from the serpent to the devil who is behind the serpent's actions. God said, "And I will put enmity between you and the woman, and between your seed and her seed; He shall bruise you on the head, And you shall bruise Him on the heel." In Isaiah 14 and Ezekiel 28, the prophets address the king of Babylon and the king of Tyre but were really addressing Satan. The same was true here. In speaking to the serpent, God was really speaking to the devil.

This first evangel, as it is often called, was God's promise to Satan, and then Adam and Eve, that He will ultimately destroy Satan and his plan to become equal to God. Scripture reveals by later prophecies that God will accomplish this through His Messiah (anointed one, Christ). A wound to the heel is a temporary wound. A wound to the head is a fatal wound. The reference to the head of the serpent would directly apply to Satan as the head of his own diabolical plan.

The seed of the woman would apply to the Messiah (Christ) and the good kingdom He would establish that will ultimately defeat Satan on the cross. He will also be defeated through all those who will believe in the Messiah's finished work of the cross.

THE VIRGIN BIRTH

The seed of the woman is an important concept. It is an important reference to the virgin birth. Notice that the death blow to the devil will come from the seed of the woman only, not man. Why is this important? Jesus Christ had to be born sinless if He was to be the perfect sacrifice for the sin of mankind. If He was born like any other man, He would inherit Adam's fallen nature and would have to die for his own sins like every person has since the fall. Being born of the woman and as a new creative act by the Holy Spirit, as we are told in Luke 1:35, Christ would receive His humanity from His mother and His deity from God. He had to be a man in order to be qualified to die for man. He had to come from God in order to be without sin.

On the basis of this explanation, we can infer that our sin nature is a product of the union of our father and mother together. It is important to remember at this point that Adam and Eve's eyes were not open to their sinful state until after Adam ate the forbidden fruit. Remember, the two were one flesh. Their sin was not complete until Adam ate of the fruit. Coming only from the woman's seed, Jesus was still connected to Adam in that Eve's body was taken from Adam's body. All this being true, Jesus could be fully human and fully God in the same person by having a human body born of the seed of the woman and the new creative work of the Holy Spirit.

That same body, being a new creation, was made alive by the entrance of the second person of the Godhead, the Son of God referred to by John 1:1 as the Word who became flesh and dwelt amongst us.

How do we know Jesus was sinless? There is one historically documented proof. That proof is His resurrection after three days in the tomb. The gospels proclaim the virgin birth but the resurrection confirms it. Death could not hold Him because scripture tells us that the wages of sin is death and the soul that sins it shall die. By coming out of that tomb, Jesus established His sinlessness as a historical fact forever. It is no wonder the devil came up with Gnostic stories, many years after the resurrection, trying to say Jesus did not die on the cross. He tried to make people believe Jesus survived the cross and married Mary Magdalene and had several children through Gnostic corruptions of the gospels. Jesus supposedly moved to India and became a guru there according to some Gnostic fictions.

Why would Satan want to produce those lies through deceived souls? An un-dead Jesus is an un-resurrected Jesus. An un-resurrected Jesus means the Old Testament prophecies concerning the Messiah were meaningless. An un-resurrected Jesus means God's redemptive work was not needed. An un-resurrected Jesus means Jesus is irrelevant and nothing more than another man. An un-resurrected Jesus allows Satan to steal the opportunity for many to miss having faith leading to salvation. But Jesus died, was buried, and rose again on the third day and was witnessed by over 500 people to be seen alive (I Corinthians 15:6).

The interesting thing about Jesus Christ is He was much like Adam before he sinned. Jesus was sinless, and yet, had choice.

The Bible tells us that He was tempted just like we are and yet He never gave in to sin. Satan came and tempted Jesus after He had fasted forty days in the wilderness. That tells us that Jesus was tested just like Adam was tested but the big difference was, Jesus overcame temptation where Adam failed. The wonderful thing is that, just as in Adam we all have sinned, if we are in Christ Jesus, we are declared righteous (forgiven). Just as we inherited Adam's fallen nature by birth, we inherit Jesus' righteousness by the new birth. It is obvious Jesus had choice, just as Adam did, but thank God, unlike Adam, Jesus chose to obey His Father. Unlike Adam, Jesus trusted His Father and obeyed His every command with a whole heart. This fact was confirmed by the resurrection. Death could not hold Jesus because Jesus was without sin.

THE WOMAN

We have seen in the confrontation that God deals with Adam first and then Eve. Here, as God pronounces the curse against them, He follows the same pattern as the record of the fall does. The temptation started with the serpent lying to Eve. Eve listened to the temptation, ate, and then convinced Adam to eat. When the time for judgment came, God began with the serpent, then dealt with Eve, and finished with Adam. God's final word was against Adam, the one ultimately held responsible by God for his disobedience. Adam was God's major disappointment.

This whole account is a study in the justice of God. God gave the command and the consequences of disobedience in chapter 2. Here in chapter 3, we see God meeting out His judgment against each guilty party according to their part played in the rebellion.

God is a God of love, but He is also a God of justice. The fall itself and the curse of death is a testimony to the fact that God says what He means and He means what He says. Those who try to argue that God, in His love, would never send anyone to Hell needs to read this historical record again. God, in His justice, must punish sin. The ongoing curse of death and the horror of the cross of Christ are grim reminders of this fact. God is love, but He is also holy. We need to remember that. Adam and Eve, being holy in their perfect state, may not have fully appreciated the holiness of God, but after the fall, their separation from God made God's holiness very obvious to them. Holy means separate. Adam and Eve understood how holy God is when their sin separated them from God in character and loss of His inward presence.

The serpent has been addressed, now it was Eve's turn to accept her judgment. God said to her, "I will greatly multiply your pain in childbirth, in pain you shall bring forth children; yet your desire shall be for your husband, and he shall rule over you." The two major components of Eve's curse were multiplied pain in child birth and coming under the rule of her husband. First, it is a paradox of life that child birth is highly painful and yet women desire to have children. This is important for the continuation of humanity's existence on earth. They were still to be fruitful and multiply but now it would be through suffering and pain.

An interesting side is we know mammals do not experience pain while giving birth like humans unless there are complications. This may be the result of the reduction of endogenously produced endorphins in humans after the curse. The endorphins work much like morphine for the animals.

This is speculation but still something to consider. Birthing for animals was not cursed.

It is also important to see that it is not a sign of weakness for a woman to want to get married, have children, and be a wife and mother. Actually, that is the kind of lifestyle by which God intended for women to find their deepest fulfillment. It is a concern today to see women activist groups indoctrinating girls and young women to deny their own God given inner longings.

Galatians 3:28 answers this demand for equality for women with men by saying, "...there is neither male nor female; for you are all one in Christ Jesus." In Christ we are all restored to our pre-fall status. Men and women are not equal in their physical make-up but they are equal in their standing with God in Christ.

Secondly, we made the point in the study of the creation of Adam and Eve in Genesis 1 that they were created co-equal. Here at the fall we see where Eve was to be under the rule of her husband. Probably the better word is protection rather than rule. God knew that the man and woman were going to be forced out of the Garden into a semi-hostile environment. The woman would be the physically weaker of the two, plus the bearer and caretaker of the children. Adam would have to be the provider of food and protector against any threats to his family. God is a God of servanthood, not domination, and Adam's relation to Eve was to be grounded in service.

Eve was being told here that the best place for her to be, after the creation coming under God's curse, was under the protection of her husband.

That is not weakness that is how God intended it to be for the woman. Men must respect this fact and not take advantage of the woman God has given to him by lording over her. A wife is to be seen as a gift, a blessing, and a responsibility from God. The curse did not change God's original intent for the man and woman to depend on each other in meaningful relationship.

THE MAN

After addressing Eve, God then turned to Adam and said, "Because you have listened to the voice of your wife, and have eaten from the tree about which I commanded, saying, 'You shall not eat from it...'"

This introduction takes Adam back to that first day in the garden when it was just him and God. Adam was newly created and in perfect harmony with God and His creation. Knowing what he did back then and experiencing the guilt and shame he was experiencing at that moment, Adam must have longed in his heart to turn back time and have another try at getting it right. But that is not how life works. We have to make the most of our mistakes and move forward. I am sure that, in the years to come, Adam would look back and remember what was lost when he disobeyed God. Fortunately, he had the hope of God's promise that God would one day destroy the works of the devil.

There is a principle to be found in what God has said to Adam about listening to the voice of his wife rather than God. Anytime someone tries to get you to do something, no matter how close you are to that person or how much you respect that person, if what they are saying is contrary to what you know God has told you to do, listen to God, not man.

We must recognize the voice of Satan speaking through that person just as Adam should have recognized Satan speaking through the voice of Eve. Jesus was tempted by Satan, and Jesus reminded him what God had said in His word. Peter tried to tell Jesus he should never have to go to the cross but Jesus recognized that was the devil speaking, not Peter, causing Jesus to reply, "get thee behind me Satan" (Matthew 16:23). Jesus overcame temptation by knowing God's commands. He knew what God's will was for His life and He stayed within that will all of His life. We must do the same.

In this introduction of the coming curse, God also reminded Adam of what he had done wrong, just as a judge does before sentencing the guilty party.

This is important to remember as a parent or person in any kind of authority. Before giving punishment, you must be able to communicate to the offender what they have done wrong. If you are not clear on what the offense was, then you do not have the right to punish. If you did not make clear what the rules were before they were broken, then you can only warn of future reprisal. In this situation, Adam knew full well what the command was and sinned anyway. God, then, was fully justified in His judgment.

As we analyze the curse pronounced on Adam, we see it had broader ramifications than those pronounced against the serpent or Eve. The curse on the serpent applied specifically to the serpent and Satan. The curse on Eve applied specifically to women. But the curse on Adam, not only applied to him, but affected all of creation. Here is what God said, "Cursed is the ground because of you; in toil you shall eat of it all the days of your life."

The curse on the ground continues to impact all of us. Apparently there was a change that took place that has affected all of mankind's relationship to working the soil. What was once in perfect harmony and cooperation with the process of growing food from the ground, no longer works the way it once did. That means there had to have been a change in the way the chemical make up of the ground reacts to seeds and the growing process. Instead of things happening spontaneously, as they would have before the fall, they would require manipulating the ground to make it responsive to man's attempts at cultivation.

In the next verse, verse 18, God goes on to say, "...both thorns and thistles it shall grow for you...," suggesting that part of the problem would be that negative plants (thorns and thistles) would grow naturally in competition with eatable plants. Those negative plants would constantly have to be removed to give eatable plants ground space. Man, then, would be in a constant battle against these changed plants that God transformed to work against man. They would no longer be the blessing God originally intended them to be. Let me make it clear that these thorns and thistles were not a new creation. They were transformed from their original structure. For example, roses did not have thorns until the fall and weeds were once a positive part of the ecosystem, not a negative. I would also add that this curse of some plants would include poisonous plants.

The word, "toil" or "sorrow," was then used by God to describe man's work in His creation. In Genesis 2:15 it says that God placed the man in the garden, "to cultivate and keep it." Man was not created to lie around and do nothing.

Work is not bad; work is good, and Adam's work before the fall was more like an avocation rather than a toilsome job. After the fall, work became something done in sorrow. A sorrow borne by the fact that man was once in complete harmony with the soil and vegetation now cursed. This would be a sorrow caused by the understanding that man's hard work just to make a living was not a part of God's original creation. Those willing to do hard work though, actually show their submission to God's will for our lives as a part of His redemptive plan for us. In light of this verse, we can see why Paul commanded in I-Thessalonians 4:11, "...work with your hands..," and II Thessalonians 3:10, "...if anyone will not work, neither let him eat." In verse 13 of that same chapter he wrote about the work ethic, "But as for you brethren, do not grow weary in well doing." Doing an honest days work is included as "well doing" and is pleasing to God.

In chapter 3:18 and 19, God goes on to say, "...and you shall eat the plants of the field; by the sweat of your face you shall eat bread..." Verse 18 goes on to say that man would now eat plants of the field rather than the fruit of the trees in the garden. This statement takes us back to Genesis 2:5 that tells us no shrub of the field and no plant of the field had sprouted. The same verse defines what it meant by those statements by saying, "...and there was no man to cultivate the ground. "This tells us that Genesis 2 was written after the fall because we find in the curse of man that cultivation of plants and shrubs in the field was a part of the curse. We need to remember that these verses were written to a people at a time when technology was very limited and most farm work was done by hand. Cultivation was hard backbreaking work and caused men and women to toil and sweat, just to put food on the table. It truly was a curse.

The eating of bread by the sweat of the brow brings up two Biblical symbols that remind us of the fall. The Passover feast included eating unleavened bread that later came to symbolize the broken body of Christ. Man's body came from the ground. The ground was then cursed but man had to live by working the cursed ground to produce the grain for his bread, the source of energy to sustain physical life. It seems reasonable to say that the Lord's Supper that uses the symbol of broken bread as the broken body of Christ, in a secondary way, tells us His broken body also broke the effects of the curse, symbolized by the eating of bread rather than fruit after the fall.

The sweat of the brow is also a symbol of man's fallen state. In Exodus 28:39-43, the high priest and the other priests were commanded to wear garments made of linen rather than wool when they entered the Holy Place and Holy of Holies.

Ezekiel 44:18 tells us why, "Linen turbans shall be on their heads, and linen undergarments shall be on their loins; they shall not gird themselves with anything that makes them sweat." The priests were to be holy when they came to minister before God. Sweat was a reminder of the curse that man fell under because of His sin against God. The keeping of the priest from sweating before God was a reminder to the priest of their fallen state before God. They needed to remember that they were allowed to approach God only because of His great mercy.

The priests not sweating before God also prophesied that God would one day do away with the curse of toiling labor. We will work in the new heaven and earth, just as Adam did before the fall, but not by the sweat of our brow.

The final element to God's curse on the man was death itself. They had already died spiritually but, in time, they would also die physically. This pronouncement of death made it very clear that, had Adam and Eve not sinned, they would never have had to die. Death was not to be a part of God's original plan. God tells Adam that He will eat bread by the sweat of his face, "...till you return to the ground, because from it you were taken; for you are dust, and to dust you shall return."

This verse gives a twinge of hope. If God has taken us from the ground by creation, He can take us from the ground, once again, by resurrection, which He has promised to do through the resurrection of Jesus Christ, His Son. The pronouncement was also another example of the reliability of God's word to know that Scripture has been telling us from the beginning that man and rocks are made of the same elements. As has been stated before, it has only been in the last hundred years that this fact has been proven scientifically, which is another example of man's science finally catching up with God's word.

THE SECOND LAW OF THERMODYNAMICS

The curse of death God pronounced on the man impacted the creation universally. We have already seen that God established the first law of thermodynamics at the end of the sixth day of creation. This law, known as the law of conservation, established the fact that God is no longer doing creation work. After the sixth day, God began to conserve what He had created. His creative work was finished and so God entered into a maintaining mode.

God conserving the universe leads us to believe that God continued to sustain the level of all processes in the universe at the optimum level they were created to operate at. He could have done so for eternity.

Then the curse of death came after man sinned. Not only did man die spiritually and began to die physically, the whole universe and all that is in it began to die with him. The curse that applied to man applied to the whole creation as well. The creation was created for man and so whatever was true of man became true of his environment. This dying of the universe has been identified by science as the second law of thermodynamics, or the law of entropy. This law describes the observable fact that all processes in the universe are running down. In other words, the universe is dying a slow energy death.

A question often asked is, "If the earth is only a few thousand years old, then why do so many of the elements have half lives that are, in some cases, millions of years long.?" My personal answer to that question has two parts. First, the creation, including the elements, was created to last for eternity. The only reason why elements have unstable isotope life at all is because of the fall and the introduction of the law of entropy. Second, even though God cursed the creation, He intended for it to continue until He completed His redemptive work. He established the decay rates of the isotopes of elements after the fall to be slow enough to allow the time needed to finish His purposes before He finally destroys this creation once for all with a fervent heat, as II Peter 3:10 prophecies.

The law of entropy is a thorn in the side of evolution theory. Evolution describes a universe that has always been progressively moving upward; continually gaining more complexity and diversity.

Entropy tells us that every thing is moving downward from a once perfect level of function. Anything left to itself will eventually begin to crumble and fall apart. The study of radioactive dating is a study of elements decaying into daughter elements. On a molecular level, metal rusts, plastic hardens, paint fades, fabrics weaken, physical bodies wear out, on and on I could go. God warned Adam in Genesis 2:17, "in dying you shall die." Micro-biology has come to show this is the case. With each new generation we produce as humans we are becoming less and less able to resist disease in our molecular make-up. Studies show our metabolism and resistance was once operating at a perfect maximum level but has been in a continual decline, as our DNA continues to lose information that makes us resistant to disease. In dying we are dying.

Dr. J.C. Sanford is a specialist in the study of the genome. In his book, *Genetic Entropy & the Mystery of the Genome*, Dr. Sanford establishes that mutations and natural selection in the passing on of genetic information have no real impact on the degeneration of that information.

He shows clearly that it is impossible for the genome to ever have moved in a positive direction by adding new information. Mutation and natural selection in the genome is moving downward in loss of information and thus cannot explain the evolution concept of genetic information moving upward from simple to more and more complex forms with increasing information.

All the information in each separate individual genetic type had to have been present from the beginning of that genetic type's coming into existence, namely the creation.

He summarizes his conclusions with the following statement, "What is the mystery of the genome? Its very existence is its mystery. Information and complexity which surpass human understanding are programmed into a space smaller than an invisible speck of dust. Mutation selection cannot even begin to explain this. It should be very clear that the genome could not have arisen spontaneously."[34] The fact is the genome in all living cells were once perfect and are now degenerating in information with fixed processes that will not allow that degeneration to reverse itself in any way, just as the law of entropy demands.

One of the best examples of entropy that is predicted by the creation model is the exponentially deteriorating electromagnetic field. We have already seen that studies show the electromagnetic field is losing strength at the rate of a 1400 year half life. We also know that, if we extrapolate back more than 10,000 years, the field would be too strong for life to exist on the earth. The creation model says that God created in six days, 6,000 years ago. It also says that God is no longer creating, but after the creation week He began to maintain His creation at the perfect level He established that would sustain the highest quality of life.

In Genesis 3, the creation model tells us God is now allowing the universe to slowly die. The creation model predicts that the electromagnetic field would be operating at its most perfect level no more than 6,000 years ago and now is weakening. That is exactly what we find.

Actually, the law of entropy is predicted by the creation model itself. If God created a perfect universe and then cursed it, you would expect to find the universe involved in a dying process.

Evolution's concept of uniformitarianism states basically that the present is the key to the past and all physical processes have been operating as they do now for billions of years. It theorizes that the way things are now are the way they have always been but increasing in complexity. The second law of thermodynamics constantly gets in the way of that theory. Just as the electromagnetic field is dying too fast, the gravitational hold on the moon is dying too fast, the erosion of the earth's surface is moving too fast while other processes are moving far too slow such as helium and carbon 14 in the atmosphere and all kinds of minerals collecting in the oceans. Like I have already said, entropy gives evolution fits but the creation model predicts it.

FURTHER REVELATIONS

The last five verses of Genesis 3 give added information that help us to see the full devastating impact of the fall. Verse 20 says, "Now the man called his wife's name Eve, because she was the mother of all the living." It is of interest that the narrative gives us this information at this point in the account. It seems to be an insertion because it comes right after the curses while they are still in the garden before any children had been born to Eve.

It makes two important points. First, it tells us that all descendants of Adam and Eve were born after the curses and were thus under those curses just pronounced by God. Second, it makes it clear that all human beings descended from Adam and Eve. There were no other races or humanoids in existence. This will be very important to remember, when we discuss where Cain got his wife. It also makes it clear that all humans come from the same gene pool. There have been studies done that trace genetic history back to a common genetic source.

That source would be Noah and his wife, as well as the genes from the wives of Noah's sons. The wives were probably fairly closely related from the family heads listed in Genesis 5. Be that as it may, the Bible makes it clear that they came from Adam and Eve.

Genesis 3:21 is pregnant with meaning, "And the Lord God made garments of skin for Adam and his wife, and clothed them."

First, this animal was more than likely a sheep. We see in chapter 4 that Abel became a keeper of sheep, partly, to use for sacrifices. Sheep were also the major animal used by Israel in their sacrificial system besides bulls. Jesus is also referred to as the lamb of God, signifying His sacrificial death. Second, Adam's first introduction to the animals was when God brought them to him to name. This established Adam's dominion over the animals as their protector. Now God has to catch one and kill it in Adam and Eve's presence to cover their nakedness. The animal would have approached God totally without fear never having experienced death before. The animal had to die in behalf of Adam and Eve. They would have to die later but that animal had to die right then. Sin always has a ripple effect. Third, God, the God of life, the living God, offered the first sacrifice (induced death) in behalf of man's sin, pointing to the time when He would ultimately have to sacrifice His own Son because of man's sin.

The blood of the animal was on God's hands when in reality it should have been on the man's. Fourth, God had to do for man what he could not do for himself. Adam had not been given jurisdiction by God to take life. He could only come up with fig leaves to cover himself. God gave man leather from animal skins.

Fig leaves and bare feet were not going to cut it out in the midst of thorns and thistles. God had to provide a protective covering for them. Actually, God gave two coverings; the covering of shed blood to cover man's sin and then, God gave a covering to protect him from the fallen creation (thorns and thistles) that was cursed because of man. What a traumatic experience it must have been for Adam and Eve to see that innocent animal slaughtered because of them. Little did they know that first sacrifice was just the beginning of a history of blood shed that was to come.

After covering them, then God said, "Behold the man has become like one of Us, knowing good and evil; and now, lest he stretch out his hand, and take also of the tree of life, and eat, and live forever...." The "Us" in this verse takes us back to Genesis 1:26 where God said, " Let Us make man in Our image, according to Our likeness.." This would be referring to the Triune Godhead, not including angels. Man was made in the image of God, yet without the knowledge of good and evil. God being totally good would know good and evil from a perfect perspective of good rejecting anything that would contradict Himself. Evil has no place in Him because He is perfectly good. There is no darkness in a shining light bulb. On the other hand, Adam could only know good as long as he remained within the covering of God's goodness. Being removed from that covering, all that was left was evil, which, as we have already said, evil is the absence of the presence of God's good.

Adam still had the capacity to know what good is, but his problem was that he could not produce that Godly good from an evil state of being. There is an expression of God's grace in this passage. God does not want Adam and Eve to eat from the tree of life any longer because of their fallen state.

He did not want them to live forever separated from Himself living a miserable unfulfilled existence under the burden of sin. He would rather them die with the hope that God would be able to finally resurrect them to a perfect life once again. Death was a curse because it was not what God intended for man. But death works for the good of man, when we know that God has something better for us beyond death, if we will trust Him. Death is an expression of what Paul said in Romans 8:28, "All things (even death) work together for good to those who love the Lord..."

We need to say something about God saying, "Behold man has become like one of us, knowing good and evil." When God created man, He did not want him to know good and evil. This tells us you do not have to have evil in order to know good, as many mystical philosophies make an issue of. Adam and Eve did quite well without the knowledge of evil. This statement by God tells us something of what our existence will be like in the new heaven and new earth. Even though God will wipe the tears from our eyes (Revelation 21:4), we will still be beings that have knowledge of good and evil. That was not true of Adam and Eve before the fall. They lived in an environment that was totally good. The wonderful thing about our new existence will be that we, like God, will then know evil only from the perspective of good. We will know what we have been redeemed from and our hearts will be full of gratitude and praise because of that knowledge.

We will never again desire evil to be a part of our existence and will be overjoyed to worship God as our God, never tempted to become equal to God again. Satan will spend eternity frustrated and angry over his defeat. We will live rejoicing over being restored to our place of opportunity to bring glory to God as His perfectly redeemed bearers of His image.

Because God did not want Adam and Eve eating of the tree of life, verse 23 goes on to say, "therefore the Lord God sent (drove) him out from the garden of Eden (delight), to cultivate the ground from which he was taken." In chapter 2, God prepared a garden for Adam and Eve to enjoy and care for. Now they are forced out of the garden to fend for themselves in a hostile environment.

To survive, they would have to cultivate the ground. It seems reasonable to assume that God intended for man to have to work hard for his food in order to keep his fallen nature in check. As the old saying goes, "Idle hands are the devil's workshop." Hard work is a deterrent to sin. Part of the problem with our society becoming more and more big city oriented is there is less for kids to do. When kids grew up on the farm, there was a lot to be done that kept them busy. One of the best things a parent can do is teach their kids how to work in order to keep them out of trouble. Here again, a curse turns out to be a blessing, if accepted and obeyed.

Finally, verse 24 brings the account of the curse to an end with very harsh words, "So He drove the man out..." What was intended for good by God, a place of delight, became a place of conflict. God had to drive Adam and Eve out of the garden because they did not want to leave. I cannot help but believe that this act of harsh judgment hurt God more than it did Adam and Eve.

God did not enjoy forcing them out of the garden any more than He enjoys sending anyone to Hell. The garden had been their home for some time. They loved being there and now God had to force them to leave because of their sin.

It is a reminder of when God had to drive Israel from His presence out of the Promised Land to go into Babylonian exile because of their sin.

It also foreshadows the time when the Son of God had to force the money changers out of the temple because they had turned what was to be a house of prayer into a den of thieves. I believe it broke God's heart, but also lead to Adam's eventual repentance. The punishment of God is always given to lead to repentance and restoration. They were never allowed back into the garden but they would still have access to God when they turned to Him in repentance.

The rest of verse 24 says, "...and at the east of the garden of Eden He stationed the cherubim, and the flaming sword which turned every direction, to guard the way to the tree of life." We do not know much about the flaming sword, except that God's word is a sword.

There are several mentions of flaming fire that refer to angels or the presence of God. One example was the pillar of fire that God used to lead the Israelites out of Egypt to the Promised Land. The cherubim were angels that were always mentioned in close relationship to the throne of God. Satan, as the high priest of angels, was a cherub who guarded the throne of God.

Ezekiel 1 pictures the cherubim carrying the throne of God as is stated in Ezekiel 1:22-26. All this suggests the guarded entrance to Eden was the place where man could come to approach God's throne in worship. As the Tabernacle was to Israel and then the Temple, so was the entrance to Eden to the pre-flood followers of Creator God.

This is even more evident when we remember the entrances to the Tabernacle and Temple were on the east side, facing the rising of the sun. The presence of God was available to those who would bring their sacrifices and find forgiveness for their sins. This will be made clear in the next chapter.

Suffice it to say, God was still accessible to those who desired to seek Him.

CHAPTER 12
WHICH
HUMAN HISTORY?

Adam's record of events after the creation now takes us several years ahead to an event that came to define the reality of the complete fallen state of the descendants of Adam, Cain murdering Abel. It would be the first great heartbreak of Adam and Eve's after their expulsion from the garden. It was a grim reminder that sin permeates mankind's fallen adamic human nature and, if not held in check, can do the most diabolical acts that can be conceived by a sinful heart. We have seen this to be true over and over in history. Men who satisfied their own selfish appetites at the expense of those more noble than themselves and, thus more vulnerable.

The post fall events of human history begin in Genesis 4. This account makes it clear that man did not start out as ape like animals that finally evolved into social animals tens of thousands of years ago. Man was religious from the beginning; he domesticated animals, built cities, developed metallurgy, and invented music instruments very early in his history. Man, before the flood, must have had a high level of technology because of what we have learned about very early societies that came into existence not long after the flood. Much of the foundations for that technology had to have been carried over by Noah and his family that quickly led to building cities of brick and mortar with towers and pyramids after the flood.

When we look later at the building of the ark by Noah and its structure being able to withstand the violent waters of the flood, we have to conclude that Noah was no slouch when it came to knowing how to build things. Pre-flood man was definitely highly intelligent from the beginning.

ACCEPTABLE

The first part of this chapter reveals what is acceptable to God in worship and what is not. In the first two verses of chapter 4 we learn important information that gives us background to this important historical event. It is important because it teaches us about ourselves as fallen beings and it helps us to see why our relationship with God is the way it is, after the fall. First, we learn that no children were born to Adam and Eve until after their removal from the garden. That means all of mankind are under the curse of sin. Second, Cain was the first born and Abel was the second born or younger brother. Eve names her first born son in response to her giving thanks to the Lord,

Who helped her to get a son. Cain probably means "gotten" or "acquired" and is seen as a gift or blessing of God, Who helped her through the birth process; something she would have done alone, unless Adam helped. It could be that she may have thought Cain was the promised seed of Genesis 3:15. We do see in this acknowledgment of the Lord that Eve continued in relationship with the Lord after the garden. This is evident in the fact that her children brought offerings to God according to her and Adam's influence. Remember, they did have the creation story available to them from the beginning. Third, we learn that Cain became a tiller of the ground, but Abel became a keeper of sheep.

There is a contrast here going back to the curse and first sacrifice in the last chapter.

Cain chose to live his life according to the curse of the ground. Abel chose to live his life according to the sacrifice by God to cover Adam and Eve's nakedness by raising sheep. A sheep was probably the animal God used to clothe Adam and Eve with; a point I have already made. Fourth, parents can bring their children up to worship God but each person is a free moral agent for better or worse, as we will see.

After the two boys were grown men, Cain was the first to bring an offering to God. Abel brought an offering to God as well. On the basis of the evidence presented in the last chapter, it seems reasonable to assume that they brought their offerings to God at the entrance of the garden much like the Israelites would bring their offerings to the entrance of the tabernacle to offer their sacrifices to God. Cain brought his offering from the fruit of his labor tilling the ground. Abel brought his offering from the firstlings of his flock. The context suggests there was more than one lamb that he brought. This account leads us to believe that some of the laws of sacrifices and offerings in Exodus, Leviticus, and Numbers trace their roots all the way back to the beginning.

There may be the possibility that each of the two sons brought a tenth of their crops and flocks long before it was required later in the books of the law. It is obvious this was a practice they learned from their parents growing up. It is no surprise that we find similar sacrificial laws in other very early (Sumer, Babylon, Assyria) cultures besides Israel's if all those cultures had their roots in Noah who followed the teachings of Adam as his direct descendent and spiritual teacher.

The importance of this incident is found in the Lord God's response to the two men's offerings.

Verse 4 and 5 say, "And the Lord had regard for Abel and for his offering; but for Cain and for his offering He had no regard. So Cain became very angry and his countenance fell." There are a few insights we need to make in response to this situation presented here. First, Cain was very happy to bring his offering. He was proud of what his work had produced and understood that the bounty of his crops was a blessing from God. We know all this because the text tells us that Cain's countenance had fallen from being happy and exited to anger and frustration, after God rejected his offering. Second, God had regard for Abel and not Cain because of their offerings, not because God loved Abel more. God is no respecter of persons. God honors the obedience of those who follow His ways. He will tell Cain in verse 7 if he does well by bringing the acceptable offering he will be accepted just like Abel was.

God rejected Cain because he brought the wrong offering, not because God loved Abel more than Cain. Third, we all have besetting sins and it seems that Cain's problem was with uncontrolled anger. Fourth, I would also suggest that Cain was angry because he felt his offering was just as good as Abel's, probably better, because he would have had to have worked harder in the fields than Abel would have raising sheep.

The point here is not about hard work though, it is about bringing the right offering. In our fallen state, we can never be acceptable to God on the basis of our work. Our work is a result of our sin and God's curse. God requires the shedding of blood to cover our sins.

Apparently, Cain was not willing to accept the requirement. Cain wanted God to accept him on his terms not God's terms. The same is true today.

People today want to come to God through self righteous works and other religions, when God says salvation through the shed blood of Jesus Christ is the only sacrifice that is acceptable to Him.

Genesis 4:6 and 7 give God's response to Cain's anger with, "Why are you angry? And why has your countenance fallen? If you do well (bring the acceptable offering), will not your countenance be lifted up? And if you do not do well, sin is crouching (like a roaring lion ready to attack) at the door (of your heart); and its desire is for you, but you must master it (before it masters you)..." (The parenthesizes are mine). Again, we see several points to be made in these verses.

First, God came and spoke to Cain just as he did to Adam in the garden after he had sinned. God was still actively involved with man after the fall.

Second, as with Adam, God asked Cain questions that God knew the answers to but Cain needed to deal with them himself. Cain needed to admit to himself that his reaction was not justified, if his main desire was to please God, not to out do his brother.

Third, God warned Cain that he was close to being overtaken by sin if he did not turn loose of his anger. Jesus made this same point in Matthew 5:22 that uncontrolled anger is as bad as murder because anger is the cause of murder, as we will see later.

There is a parallel to be made here. Just as sin crouched at the door of Cain's heart in this dilemma, Revelation 3:20 has the Lord telling us He stands at the door of our hearts and knocks. God was actually knocking on Cain's heart's door calling him to repentance for his unjustified anger before it turned to murder.

Fourth, God told Cain he was responsible to master his anger before it took control of him. Adam and Eve must not have helped Cain much with controlling his temper.

Being the first parents was not easy for Adam and Eve. They had no previous examples of raising children before them. They learned with their first son that sparing the rod spoiled their child. Cain's actions certainly suggest he was a spoiled child. Parenting is about helping our children master their potential besetting sins before they get out of control as the child gets older. Parents can work with their children by not allowing temper tantrums, helping them discipline themselves to do what they are told, talking to them about coming temptations of drugs, alcohol, and sex before they are confronted with these temptations.

Parents should also be monitoring what their children watch on television or look up on the computer. Now days, parents need to check out their local public and school libraries. They often have books available to kids in the children's sections that they are not ready for. They must know exactly who their children's friends are and what kind of parents they have. You can never be too cautious. As a pastor, I had a family in my small town church whose twelve year old daughter was being molested by her best friends parents when she spent the night at her friends house. You can never be too protective of your children in this day in time.

CHAPTER 12 - WHICH HUMAN HISTORY?

Hebrews speaks of sin that, "so easily besets us" in Hebrews 12:1. The idea of "beset" in the Greek has to do with sin that stands easily around us or surrounds us as a means of distracting us or diverting our attention. What was inferred in that passage was, because we are surrounded by so many witnesses to the faithfulness of God in overcoming sin, we should be encouraged to break those enticing entanglements of sin as well. We should no longer allow it to distract us. To do so, we must recognize that sin for what it is and, through Christ, break away from its enticements.

We all have different sins that entangle us more than others. Cain's besetting sin was an uncontrolled temper. He did not listen to God's warning that his anger was sin that must be controlled, and so, it leads to deeper sin. For others, a besetting sin may be an addiction to alcohol, drugs, pornography, eating, jealousy, envy, recognition, power, etc. Some become controlled by fornication, lying, cheating, stealing, or rebellion to authority.

In this day in time, the gay and lesbian life style is being justified by claiming that those who have adopted this lifestyle were born that way. The problem is that God's word says it is sin because it violates God's original standard for one man to have one woman as a partner for life. It would be unjust for God to declare something a sin and then turn around and make someone that way in the womb. The problem is not that God made us to be a certain way. The problem is that we are fallen in sin and homosexuality, like all other sin, can beset those who are vulnerable to that temptation or enticement. But God says that we must master those temptations, before they take control of us. If you do not master sin, sin will master you.

Homosexuality is a learned pattern of self identification. As we develop from babies into adults, we are influenced by all kinds of stimuli that suggest positive or negative impressions of ourselves. Children are not sexually aware until they reach puberty, normally. Now days, because of all the bombardment of sex in our society, that is becoming less true. But normally, children naturally develop close bonds of friendship and love with those of the same sex, until their attraction to the opposite sex begins to grow as they become teens. Some children may see this love for their same sex friends, as an indicator of being homosexual.

If affirmed in that impression by significant others, they may learn to resist their natural development of attraction to the opposite sex. Some boys are effeminate, and some girls are tom-boys. If these children are not taught that there is nothing unnatural with boys liking girl things and girls liking boy things, then they will be vulnerable to being type cast by their peers as gay when made fun of. Children are very good at attacking other children at their weakest points. When you are called gay enough times, after a while you start believing it must be true. A lie repeated often enough becomes true to those who have been programmed to believe it. People who are vulnerable to this scheme of the devil to destroy them must recognize it for what it is and learn to resist the temptation of this besetting sin in the power of the Holy Spirit.

The real issue here is that Cain was not able to master his sin because he was fallen and out of fellowship with God. He would not admit to himself his sin problem. No one ever begins to have victory over any kind of sin until he admits to himself and agrees with God that he has a problem. Was Abel perfect?

No he was not; but he was declared righteous because he brought the acceptable sacrifice that put him in right relationship with God. Cain could have been accepted as well, if he would have done what his brother had done. The message of the whole Old Testament is that man, in his own strength, is not able to overcome sin. That is why we need a Savior to cleanse us of our sins. Then it becomes possible for God's Holy Spirit to live in us and to empower us to overcome sin by His presence living in us. The old things (sinful nature) begin to pass away so the new can come (the new man in Christ).

Chapter 4:8 records the first great tragedy that took place after the fall, "And it came about when they were in the field, that Cain rose up against his brother and killed him." The structure of this verse suggests that Cain related to Abel what God had said. It may be that Abel may have encouraged Cain to do what God told him to do. Abel would soon learn that darkness hates the light when those in the light tell those in darkness what they need to hear. Several Old Testament prophets lost their lives for exposing the sins of godless kings. John the Baptist was one of them.

There is a period after this first sentence telling us that the Hebrew implies this is an event unto itself. It causes us to assume that Cain stewed on God's rejection of his offering to the point that he finally took his anger out on his brother. The fact that Cain killed Abel in the field rather than in a pasture suggests Abel had come to Cain where he worked in the field, rather than Cain coming to where Abel was working with his sheep in the pasture. It could be that Abel came to Cain where he was working in the field and tried to entreat his brother once again to listen to God. Regardless of how the two men came together, it was more than Cain could stand, so he killed his brother.

There are two pictures which we need to see here. First, instead of obeying God and bringing the acceptable sacrifice, Cain selfishly sacrificed his own brother, Abel, not being willing to accept God's required offering. We do the same with Jesus. We do not obey God so we sacrifice Jesus because of our own rebellion. Like Abel became Cain's sacrificial lamb, Jesus Christ became our sacrificial lamb. Second, There is the picture of the older brother slaying the younger brother because the younger brother was more righteous (Hebrews 11:4) than the older brother.

This picture is a prophecy of the descendants of the first Adam (the older) sacrificing the second Adam (I Corinthians 15:45) or younger brother who was the more righteous one. It is amazing to recognize that in Genesis 3:15 God promised to deliver man through the seed of the woman. In the very next chapter, God's word gives the prophetic picture of how the seed of the woman would crush the head of the serpent by being sacrificed by the hand of His older brother (Adam's descendants). This is another example of the Biblical worldview being consistent with itself as we see the prophetic pictures of the Old Testament being finalized in the New Testament.

ANOTHER CONFRONTATION

Just as we saw in Genesis 3, here in Genesis 4 God came to confront Cain with his sin in the same way He did his parents after they had sinned. This confrontation reminds us that God is just and must deal with sin. Here again, as we asked in chapter 3, why did not God stop Cain before he killed Abel? God knew that Cain's anger could lead to murder because He warned Cain about that anger.

We see God intervene in the story where He told Abraham, in Genesis 22, to take Isaac to mount Moriah and sacrifice him. In that circumstance, the angel of the Lord told Abraham to stop before he plunged his knife into Isaac's heart. God did step in there. Why? Because Abraham was acting in obedience to God's command. God did not violate Abraham's free moral agency in this case. First, Cain, like his parents, was a free moral agent and had been warned by God before Cain chose to kill his brother. He was still free to choose to obey or disobey God. Second, God was working out His purpose through history.

It is in His purpose that some live and some die at the hands of evil men according to His purpose. That does not mean the ones who die are less blessed of God than those who are delivered from death by God. God did not deliver Jesus from the cross because it was His will for Jesus to die in order to accomplish God's purpose. We see this in the lives of the two disciples, James and Peter, and Stephen the deacon. James was killed by Herod very soon after Pentecost, as was Stephen. Peter, on the other hand, was delivered miraculously by the hand of God when Herod was going to kill him, after killing James.

The church prayed for all three men but only Peter was spared. Why? God accomplished more through the deaths of James and Stephen than had they lived. God accomplished more through Peter's life than had he died early in his ministry. Remember, Peter was martyred later. Abel being mentioned in Hebrews 11:4 as the first martyr of faith tells us Abel did not die in vain. Instead, he became an example of all those who obey God who will have to make personal sacrifices for God at the hand of Godless men.

Genesis 4:9 begins God's confrontation with Cain by His asking the question, "Where is your brother?" In chapter 3, God asked Adam, "Where are you?" At this time He has to ask Adam's son, "Where is your brother?" The question asked of Adam reminds us of our responsibility for knowing where we are in our relationship with God. The question asked of Cain reminds us of our responsibility for our fellow man. God does not just ask, "Where is Abel?" Instead, God puts the emphasis on the fact that Abel was Cain's brother. This brings home to Cain the realization that Abel was not a stranger or an enemy but his brother. Abel was a person he was supposed to love more than himself.

The law would teach later, "Love thy neighbor as thy self." In Adam, our neighbor is our brother or sister. We are all blood relatives.

God, by asking the question in this way, established the fact that He expects us to know the circumstances of those close to us and to help them when needed. This was established before government ever came into existence. Government is not God's chosen avenue for man to care for man. Government requires forced revenue through taxes to maintain law and order. We are responsible to God as individuals to care for our brother out of our own free will.

Government welfare is an unjust transfer of income from one group of citizens to another, usually against the taxpayer's will. We are all brothers in Adam. It is the people of God who should be effectively expressing God's care for everyone by helping those in need out of their own free will. There is more than enough money in the churches of the United States to do away with welfare.

If the government would stop stealing welfare money through taxes and if the churches took up the responsibility for taking care of the poor like they should have been doing all along, government welfare could be vastly reduced.

This question, "Where is your brother Abel?" is also the foundation of the seventh commandment, "Thou shalt not murder." This commandment distinguishes between the murder of an innocent person and the need for war or capitol punishment to stand against those who do not respect life. The Biblical worldview teaches we are responsible for each other and the willful taking of innocent life is not acceptable. It stands in stark contrast to evolution that glorifies the survival of the fittest and dominance of the strong over the weak.

Evolution theory, when taken as it was originally stated by Darwin through survival of the fittest, glorifies gaining the upper hand over the "weak" at any price. Only the strong should be allowed to survive. Evolution theory provides no basis for expressing compassion of any kind.

Cain's response to God's question, "Where is your brother Abel?" shows a total lack of remorse or willingness to repent on Cain's part. He answers with a lie, "I do not know;" and with self-justification, "Am I my brother's keeper?" People unwilling to answer a threatening question will often answer with a question, as Cain did. Over the centuries, God's answer to Cain's question has been a resounding, "Yes, you are your brother's keeper." Taking care of the poor, widows, orphans, and strangers is a common theme throughout scripture.

In Matthew 25:31-46, Jesus warns of the coming judgment where He will separate the sheep from the goats on the basis of how each individual took care of the needs of others. God is very concerned about how we take care of our fellow man. In the life of Christ, we see that most of His miracles were done to meet needs and relieve suffering. He did not turn rocks into gold or raisins into rubies to enrich Himself. He fed the hungry and healed the sick. The best way for the people of God to bring the lost into the kingdom of God is to meet needs and then tell the story of God's love through Jesus Christ. Sharing the gospel without practical expressions of God's compassion is like sounding gongs and tinkling cymbals in the ears of those we may try to win. As we relieve suffering, it gives us the right to share why we care.

Then came the resounding question full of God's wrath and indignation, "What have you done?"

What is the answer to God's question? The following are four statements of what Cain had done.

First, Cain had shed innocent blood as a result of his uncontrolled anger.

Second, he had sacrificed his brother rather than being willing to repent and bring his proper sacrifice to God. His anger was fueled by misplaced envy and jealousy. Abel was not his problem, he was his own problem.

Third, his anger was really directed toward God more than Abel because God did not accept his offering. Cain took his anger against God out on his brother. Cain was demanding fairness, as we so often do; but God is not about fairness, He is about justice and obedience.

Life cannot be fair in a fallen world that has overlapping consequences that affect both the guilty and the innocent. Life is all about the sovereignty of God and our learning to live within that sovereignty.

Fourth, Cain fully expressed his unchecked total depravity. All four of these were what Cain had done.

The liberal understanding of man is that he is basically good and, with the right environment and proper education, will do good. The Biblical testimony about man is he is depraved and, if not held in check by belief in future rewards and punishments, man will do terrible things. A nation that is not under God is a nation that has no ultimate authority and will ultimately be ruled by the depravity of man. Every Communist, Nazi, or Fascists nation in history that tried to do away with God exhibited little if any concern for human rights or the sanctity of life. The French Revolution became a blood bath of anarchy as godless men took control of that movement. When God is removed, the true nature of man exposes itself in all its cruelty.

The next statement made by God tells us God is very concerned for those whose innocent blood is shed unjustly. God is concerned and will bring justice in His own time. God revealed the foolishness of Cain, who thought God did not know what he had done when He cried out, "The voice of your brother's blood is crying out to me from the ground." The voice of Abel's blood cried out to God because his life was in the blood. The tense of the verb suggests a continuing crying out. This is supported in Revelation 6:10 where those who had been martyred for the word of God continually cry out, "How long O Lord, holy and true, wilt thou refrain from judging and avenging our blood on those who dwell on the earth?"

Verse 11 tells them God will do so when the rest of those who will be martyred in the future have been killed also. Jesus also said in Matthew 23:35, as He warned the scribes and Pharisees of His day, "...that upon you may fall the guilt of all the righteous blood shed on earth, from the blood of righteous Abel to the blood of Zechariah, the son of Berechiah, whom you murdered between the temple and the altar." God will bring judgment on all those who reject Christ and, in doing so, murder His saints. We must tremble to think of the millions of babies and children who have been murdered by abortion or infanticide over the centuries and especially in this day where the major political concern is protecting the right for women to have abortions. The blood of all those babies, just as Abel's, continues to cry out to God for justice. God's word is consistent with itself.

ANOTHER CURSE

God is now going to pronounce another curse, just as He did against the serpent, the woman, and the man, only this time it will be against Cain.

The difference was that Cain's curse only applied to him. The sad thing that we will see was after Cain left the presence of God, his descendants will loose touch with God and finally have to be destroyed by God in the flood. His leaving the presence of God actually would become the norm rather than the exception in the years to come. This reality is a reminder that man, left to himself in his fallen state, does not seek God but follows the lust of his flesh, the lust of his eyes, and the pride of life that does not think it needs God for anything.

In Genesis 4:11 and 12 God pronounced this curse against Cain, "And now you are cursed from the ground, which opened its mouth to receive your brother's blood from your hand. When you cultivate the ground it shall no longer yield its strength to you; you shall be a vagrant and a wanderer on the earth." This was a very just sentence against a guilty and unrepentant man. First, working the ground was Cain's first love. It was that greater love for his work than willingness to obey God that caused his anger in the first place.

It put him in competition with his brother that led to hate and murder. Second, Cain was to be an example of how God uses things against us that we love more than Him. He often does so by taking it away. There were no prisons at that time, and so, Cain became God's prisoner for life. He got a life sentence of wandering rather than farming. He lost the opportunity to do what he loved most. Third, God hit Cain at the point of his pride. He was proud of the work he had done, bringing God the fruit of his labor to impress God, forgetting that God is the one who gives the increase. His curse would remind him over and over again that it is God who makes things work in His creation; man can only help it along.

Cain's response to God's curse is typical of an unrepentant person. Verse 13 replies, "My punishment is too great to bear!" Notice, Cain is only concerned about himself. What about Abel? What kind of punishment was he having to bear for doing what was right?

Depravity puts more emphases on the rights of the guilty rather than the cost inflicted on the victim. A truly repentant person will accept the consequences of his actions. One of the thieves on the cross next to Jesus admitted his guilt and received forgiveness unto eternal life.

The other thief died in his sin totally unrepentant. In this account, God heard two forms of crying; Abel's blood was crying from the ground and Cain crying because of his punishment. We see God's justice affirming the victim, Abel; and punishing the guilty, Cain. It is the opposite today. Society is so concerned about protecting the rights of the criminal that the victim is often forgotten when it is time for justice to be served.

In Genesis 4:14, Cain goes on to complain concerning his plight, "Behold Though hast driven me this day from the face of the ground; and from Thy face I shall be hidden, and I shall be a vagrant and wanderer on the earth, and it will come about that whoever finds me will kill me." First, Cain complained that, not only will he no longer be able to cultivate the ground, but he will not have access to God. As a vagrant and wanderer, Cain would no longer stay in one place long enough to develop the land, but more than that his wandering would take him away from the entrance to the garden where God could be approached. This reminds us that Cain's descendants would later have to be given the location of the garden, as we see Adam did in Genesis 2.

Second, Cain was concerned about being killed by other people who might find him. This raises the question by evolutionists who would ask, "Where did these other people come from, if Adam and Eve were the first humans and Cain and Abel where their first two born children?" There should be no one else, they reason.

To answer this valid question, we have to go no farther than Genesis 5:4 where we are told Adam had other sons and daughters. Josephus records in his footnotes that Adam had 33 sons and 23 daughters, a total of 56 children.[35]

This record only mentions the one other son, Seth, who we already know was born after the Cain and Abel tragedy. Seth is the only one mentioned because, as you read the rest of chapter 5, it lists only those men who apparently continued to worship the Creator God when others did not.

This would tell us that there may have already been other sons and daughters born to Adam and Eve before Cain killed Abel. In light of this, we can see that the avenger of blood, found in the law of Moses, goes back to the beginning; and the cities of refuge were given to keep Israel from shedding the innocent blood of those who accidentally killed someone. Revenge most certainly is a natural response of the flesh. Avengers of blood may have been one reason why there was so much violence on the earth before the flood that God finally had to put a stop to it. Cain was certainly concerned about the avenger of blood.

God's response to Cain was an encouraging one. Almighty God, in the midst of His judgment, listened to the cry of a murderer. We see in this verse that God's grace and mercy can be gained by the worst of sinners. There is hope for us all. God responded to Cain's complaint by saying, "Therefore whoever kills Cain, vengeance will be taken on him seven fold." Two wrongs do not make a right. It seems that God wanted to stop more blood shed before it got started.

Later God would invoke capital punishment, but at that time, Cain's punishment was enough. It is important to remember that these events took place very early in man's history before the flood. There would be many changes made after the flood, including capital punishment.

The rest of verse 15 speaks of the mark of Cain. Evolutionists in the past tried to use this mark as the explanation for where black people came from. There is no justification for that kind of bigotry in scripture. The scripture makes no attempt to describe this mark. What kind of mark it was is of no consequence. The point is that God gave Cain an obvious mark that, when people saw it, they would know it was Cain. We will learn in a later chapter how mankind developed different physical characteristics due to genetics, not by some curse of God. All of mankind continues to be equal by being made in the image of God regardless of physical characteristics.

Genesis 4:16 tells us that, after Cain was cursed by God, that he, "...went out from the presence of the Lord, and settled in the land of Nod (wandering), east of Eden." The key here is that Cain left the presence of the Lord. He not only left His presence physically but he also left the influence of God's presence. First, he left the land of Eden where access to the garden gate was and went east to the land of Nod. Liberal commentators try to suggest that this is another contradiction to the idea that Adam and Eve were the only humans in existence by saying this other country was already in existence at the time of Cain.

The obvious explanation is that the land where Cain went to became the land of Nod (wandering) because that was where Cain went to wander. That is how Nod got its name. This account obviously was being written after Nod became known as Nod. Second, Cain's leaving the presence of God's influence meant he chose to no longer be a seeker of the one true God. Those who choose to reject God finally become philosophical and emotional wanderers.

Truth eludes them as all things become subjective and relative lacking any foundation for meaning outside of themselves.

He went east toward the rising of the sun possibly suggesting that he became a worshiper of the sun, moon, and stars. Cain's rejection of God's influence would give rise to the age old distorted worship of the constellations later to be called the Zodiac in astrology. Cain moving east away from the presence of God brings up something worth considering.

In scripture, the East seems to symbolize the place where man moves away from God into sin. Noah migrated east from Ararat into Mesopotamia where Babylon and Nineveh were established. Abraham moved west, opposite of east, from Ur of the Chaldees to what would become the Promised Land where God had commanded him to go. The Jews were sent east to Babylon to be punished. I bring this up because the Garden gate faced the East. The tabernacle entrance faced the East. The temple entrance also faced the East.

What that says to me is that the places of God's presence faced east as a sign of God calling sinners back to Himself. Just as the wise men came from the east to worship the promised King in the person of the baby Jesus, wise men come out of their sin back to the presence of God who keeps His doors open to all those who will come.

CAIN'S DESCENDANTS

The next verse has given rise to one of the most asked questions by those who want to justify their rejection of scripture as a reliable historical record.

What is this world renowned question? "Where did Cain get his wife?" That question has already been answered by our reference to Genesis 5:4. Adam and Eve had other sons and daughters besides Cain, Abel, and Seth. Some of those children were probably already alive when Cain killed Abel. That would explain why Cain was worried about an avenger of blood coming after him from Abel's family. Cain leaving the presence of God and going east would also mean he would have to take his wife with him.

Why? Cain and his wife would become the first settlers of Nod. It is more than reasonable to assume that Cain's wife was also his sister. At that time, inner marriage amongst siblings was not only accepted it was necessary to continue the human race. We see it was a common practice, at least, up to the time of Abraham, who married his half sister, Sarah. Isaac married his cousin, Rebekah. Leah and Rachel were Jacob's cousins as well. Incest at that time was not a danger to the gene pool because genetic mutations had not yet become a threat to the health of future generations. Once that became a problem, incest became forbidden.

Genesis 4:17-22 gives a list of the descendants of Cain. Cain's son was Enoch. The first city mentioned in scripture was named after him. Enoch's son was Irad; Irad's son was Mehujael; Mehujael's son was Methushael; and Methushael's son was Lamech. These men represent five generations after Cain. It was Lamech who began the practice of polygamy and, from his two wives, Adah and Zilla, came the men who established the foundations for culturalization.

Adah gave birth to Jabal who developed living in tents and domesticated animals beyond sheep, which apparently, established the concept of accumulating wealth. His brother was named Jubal and he discovered the art of making musical instruments and playing them. Zilla's only named son was Tubal-Cain who developed the process of making implements out of bronze and iron (so much for the progression of the stone age to the bronze age to the iron age of secular human history). He had a sister named Naamah we are told.

There are some observations that need to be made about this passage of scripture. First, there is a correlation between the names of Cain's descendants and the names of the descendants of Seth found in Genesis 5. Enoch is the same as the Enoch who walked with God and a name very close to Enosh, the son of Seth. Mehujael and Methushael are similar to Methuselah who lived 969 years and was the son of Enoch. It is also interesting that both Methushael and Methuselah named their sons Lamech.

The first Lamech killed a man and the second Lamech was the father of Noah. All this suggests these names became common among humankind and there may have been a connection between Cainites and Sethites for a time. It suggests that they maintained a common language all during this time. It also suggests that some Cainites continued to worship God for a time, at least, before Lamech married two wives and killed a man. This seems to be the time when each group split and Adam lost contact with Cain's family. They must have gone their separate ways both religiously and geographically. As the human population grew, Adam and the faithful followers of the Creator slowly lost their influence in the world.

The next observation that can be made is the mention of a city being established by Cain that was named Enoch, after his son. Early in the Old Testament, the followers of the Creator God were encouraged to stay away from cities. Cities and rebellion against God seem to have a correlation. Some examples of these cities were the city of Nimrod's rebellion, Babel, and Ur that Abraham was called out of to live as a wandering shepherd in Canaan. Sodom and Gomorrah were places to be destroyed. Moses led the Israelites out of the cities of Egypt they were helping to build to go and worship God in the wilderness.

They were told to destroy all the cities of the Promised Land because of their wickedness. All this tells us that it was no coincidence that the first city was built by Cain. It was probably named Enoch because Cain, as a vagabond and wanderer, would have started it but did not stay around to build it. He left that task to Enoch. The city would be the first of its kind and would be populated by Enoch's descendants, the most prominent ones being cited in the Biblical record. The "city" seems to represent man's way of becoming self-sufficient and the environment where immorality takes hold. The Biblical message to God's people is always, "Come ye out from among them."

Cultural advancement took off with the son's of Lamech. These were not cave men grunting around living in caves trying to eke out a living. They developed the technology to make tents from skins or cloth. They also domesticated animals, suggesting a use of them for doing work. They made instruments for producing music, including wind and stringed instruments. This would require much innovation, including an understanding of mathematics. They also developed metallurgy of bronze and iron which would require some understanding of chemistry and developing molds.

This development of metal tools by pre-flood man is supported very strongly in rock strata. Iron pots, gold jewelry, iron bands, an iron cube, and other tools to name a few have been found in the deepest of strata. In London, Texas, a hammerhead was found in cretaceous rock that was made of a mixture of elements modern man has not been able to duplicate. This suggests the metal was produced in a different environment from the present one by highly intelligent people.. It is owned by the Creation Evidence Museum in Glen Rose, Texas. There is no doubt that as the population of man grew, many more cities sprang up and technology increased at a rapid rate given the high intelligence of man. We will see later that man used brick to build the tower of Babel. It is very possible that Noah and his family knew how to make brick before the flood and brought that technology with them off of the ark.

Chapter 4:23 takes an unusual turn. It introduces a poem spoken by Lamech to his two wives. Some scholars have said that this poem is very ancient and yet shows a high level of creativity in the original language. Once again, the high intelligence of early man is confirmed. It is the content of the poem that is unusual. Lamech brags to his wives about how he killed a young man. Lamech gives his boast by saying, "Adah and Zilla, Listen to my voice, You wives of Lamech, give heed to my speech, for I have killed a man for wounding me; a boy for striking me; If Cain is avenged sevenfold, Then Lamech seventy-seven fold."

This poem gives us several points to ponder. First, it tells us Lamech knew the Cain story well, though being five generations removed from Cain. If people lived longer in those days, as we will see they did in the next chapter, Cain was probably still alive at Lamech's writing of this poem. Second, Lamech seems to refer to Cain's protection as unjustified because he killed Abel without cause, whereas, his killing was justified by self defense. In a way, he is demanding from God protection from the avenger of blood because of what God did for Cain. Third, in Genesis 6 we will see where God announced to Noah His plan to destroy the earth because of the violence that had become the norm.

The violence began with Cain and increased with Lamech. It continued to increase until it had to be stopped by a drastic move of God. This poem seems to have been recorded by Adam to show the further descent of man into his depravity after the fall. Fourth, the seventy-sevenfold reminds us of how Jesus told Peter that instead of forgiving a brother who sins against us seven times we must forgive them seventy times seven. Unlike Lamech, Jesus taught us to turn the other cheek if someone strikes us and to forgive others as often as we want God to forgive us. Depravity always finds a way to justify its evil. Righteousness always finds a way to do good to all men.

FINAL TRANSITION

Adam, being finished with the account of Cain and the results of his curse, transitions back to the family situation that continued seeking God's presence close to the garden entrance. He acknowledged the grace of God that accepted the birth of another son as God's replacement of Abel.

The scripture says that Eve named their new son, Seth (knew), saying, "God has given me another offspring in place of Abel; for Cain killed him..." (Genesis 4:25). This son, Seth, being seen as the replacement for Abel indicates that he replaced Abel as the spiritual leader more than just as another son to remove the gap left in the family.

There is a hint of Eve's disappointment and frustration with Cain as her first born son in the words, "for Cain killed him." She held such high hopes for Cain and he broke her heart by killing Abel. The birth of Seth was like a birth of new hope. God in the midst of tragedy brought newness of life.

The last verse, 26, of chapter 4 says, "And to Seth, to him also a son was born; and he called his name Enosh. Then men began to call upon the name of the Lord." In contrast to Enoch, the son of Cain, who built a city, Enosh continued the worship of the Lord that apparently began with Adam and Seth. In chapter 4, we see where the name of God has been simplified to Lord rather than Lord God as it was in chapter 2 and chapter 3. This does not have to suggest a different writer as much as an on going development of man's relationship with God. It can be seen as a progression over time.

The real issue is that Seth and, then Enosh, reestablished what was lost by the death of Abel, namely the true worship of God. Cain left God's presence but Seth and then Enosh continued on with the Lord. In chapter 5, we will see that each new generation had someone to take up the worship of God. It reminds us of the importance of how we live our lives. Will we influence those who come after us for good or for evil. Seth and, then Enosh, established a pattern for good.

Genesis 5:1 begins by stating, "This is the book of the generations of Adam." We need to remember that the original text of the Bible was not written with chapter and verse divisions. Those were added later to help reference Bible passages more effectively.

If we follow the interpretation of Genesis that suggests the book is a collection of different writers who's writings end with a toledoth, then this sentence should be the final statement of chapter 4 rather than the first of chapter 5.

That would mean that Adam was the eye witness to all the events recorded previously from Genesis 2:5 to Genesis 5:1a, as the sentence says. We know that Moses was the compiler of Genesis but he got his information from written records passed down to him. Adam was the first human inspired writer of God's word. Moses was the inspired compiler of the inspired *toledah* that make up the book of Genesis.

If Adam was the source of the information in chapters 2, 3, and 4, then that would tell us something about Adam. It would tell us that he wrote his testimony as an act of repentance to explain to his heritage why there was pain, suffering, and death in a world created by a good and loving God. His point would be that God was not the enemy, but that he, Adam, was to blame for the fall. It was his lack of trust in God's word that lead to his disobedience and ultimate separation from God. But, his testimony gives hope and a promise to believe that God would finally destroy Satan's evil scheme of self deification. God was going to destroy the devil through the seed of the woman. Adam sets the record straight about why things are the way they are and does away with godless theories rooted in evolutionistic assumptions and false philosophies pertaining to God and man.

Adam's eyewitness record makes it clear that man has always had enough of God's inspired word available to him to establish faith in the one, true, Creator God.

THE LINE OF LIGHT

Genesis 5 gives a list of what I call the line of light. It begins with Adam and ends with Noah's three sons. What is the meaning of this list?

The answer is found in chapter 4 where Adam tells us that with Seth and Enosh, "...men began to call on the name of the Lord." Cain left the presence of God and started a heritage that led to God's judgment. In chapter 5 we find the list of those men who continued to call on the name of the Lord in continuation of the example made by Adam, Seth, and Enosh.

Their influence led to Noah's finding favor in the eyes of God that gave humanity another chance. Chapter 5 follows a pattern, after the introduction, that summarizes the creation of Adam and Eve in Genesis 1. This summary (verses 1and 2) makes it clear that this was the beginning of the next account. The pattern is 1, the age of the father when his son was born; 2, the statement that the father had other sons and daughters; 3, how much longer the father lived after his son was born.

There are three important exceptions to this pattern. First, the account makes the point that Seth, the son of Adam, was made in Adam's image, not God's, suggesting Seth's inheritance of his father's fallen nature.

Second, we are told that Enoch walked with God and God took him; meaning he did not die but was physically transported alive to heaven much like the prophet Elijah was. Some commentators believe these two men will be the two witnesses during the tribulation in Revelation 11 because they have not experienced death yet. Third, the three sons of Noah are named rather than just one and the age of Noah when he died is not given until chapter 9:29. This would be the case, if Noah was the compiler of these chronologies, as we see he was in Genesis 6:9. He would not have recorded his own death.

Here are some other important observations about chapter 5. First, all the men listed were not the first born nor were they the only child.

Seth was born after Cain and Abel as well as their wives and possibly some of their other sons and daughters. These men were listed because they were the ones who continued having faith in the creation account and Adam's testimony. They were the spiritual leaders of their clan in their generation. They would be considered to be God's remnant up to the flood. Apparently many of their sons and daughters did not continue to worship God after they grew up and started a family. They chose to go the way of Cain, who left the presence of God. This is evident when Noah and his family are the only ones left having faith when the flood came.

Second, all these men lived a long time compared to the post-flood ages and those of today. Methuselah lived the longest at 969 years, and Enoch lived the least at 365 years. He did not die, meaning he would have lived as long as the others if he had not been taken. That certainly does make the point that these men lived in a much different environment than we live in today.

Third, on the basis of these time tables, all the men on the list lived at the same time as Adam, except Noah, who was born 126 years after Adam's death. Noah lived at the same time as all the men on the list except Adam, Seth, Enosh, and Enoch. What that tells us is the pre-flood world had access to Adam for more than half of the time between the creation and the flood. The eye witness who saw and knew God personally or someone who knew Adam first hand and believed his testimony were alive until the flood. When Noah preached for 120 years before the flood, he had Methuselah and Lamech to back him up.

Both men knew Adam personally. The people who died in the flood were without excuse. According to Jude 14, Enoch was also a prophet in his time that preached God's judgment before Noah was born. There is some evidence that the name he gave his son, Methuselah, means "when he dies, it will come." That prophecy made through his son's name has reliability because Methuselah actually died in the year the flood came, 1656 years after the creation. Fourth, the fact that chapter 5 is written in such a way that provides a specific time table between the creation and the flood, there is no doubt God wanted it understood in His inspired word exactly how old the earth was when the flood came. The rest of the Old Testament follows a similar pattern that continues on with a commitment to keeping up with the passing of time. Man's science can say the earth is old but the Bible makes it clear that the earth is very young.

There are many indicators in the natural world that point to a young earth. The spin of the earth is slowing down at a rate of 1/1000 a second every day. A few thousand years of slowing down at that rate would not be life threatening to the planet.

But, if evolution is true, the earth would have had to have been spinning so fast 4.5 billion years ago that life would not be possible on earth, in order for it to have slowed to its present rate.

The growth of active coral reefs could not have taken more than about 10,000 years. Polystrate fossils, such as full grown trees, are found all over the earth in coal seams or embedded across what is assumed to be millions of years of strata. There is no way these fossil trees as well as other fossils could have survived being buried slowly over millions of years.

The better explanation is these fossils were buried rapidly four thousand years ago in the flood when the strata layers and coal seams were all formed. The population growth rate known for the last several hundred years tells us man has been around for about four thousand years, only as far back as the flood.

Suffice it to say that God's word and good science are on the side of believing in a young earth. Genesis 5 is the foundation of this young earth assumption.

CHAPTER 13
WHICH FLOOD?

The Biblical record is taking us on a historical journey that has now brought us to the most controversial event in history, at least from geological and theological perspectives. If the earth is old and evolution or any other worldview that holds to an old earth is true, then a worldwide catastrophic flood that covered all the mountains cannot be true. Why? That kind of catastrophe would be a prefect explanation for all the geological formations around the world. It would also confirm the validity of the Biblical worldview. This confirmation would destroy old earth theories once and for all. Secular, old earth, geologists reject the worldwide flood completely. Old-earth theologians that are influenced by secular geology try to reinterpret Genesis 6 and 7 to describe a local flood that took place only around the Black Sea or Mesopotamian Valley.

Although not popular now, some theologians have tried to say the flood was tranquil and had little effect on changing the surface of the earth. That theory became almost laughable, when simple observation tells us there is no such thing as a tranquil flood. As we go through the text of chapters 6 and 7, we will show how the text itself completely denies the possibility of any other explanation except a worldwide flood. We will also show how the worldwide geological formations are best explained by one catastrophic flood and how the fossil record is better explained by the flood than the slowly developing "old earth" geological column.

There is one point that can be made that hints at the fact that even secular geologists are coming around to a worldwide flood catastrophe. Many of those geologists have scrapped trying to explain local geological formations by the slow developing processes of uniformitarianism. Instead, they are now explaining them as local flood catastrophes. One of these days, as they keep finding these local catastrophes all over the world, they will realize that they are all better explained by one worldwide event, such as a flood.

THE FLOOD JUSTIFIED

Noah's account continues in chapter 6 to describe what was taking place on the earth that led to God's saying, "Enough is enough." The first abominable practice that Noah described was the sexual union of the sons of God and the daughters of men. The immediate question that comes to mind is, "who are these sons of God?" Those interpreters who shy away from supernatural explanations explain the sons of God to be the descendants of Seth and the daughters of men to be the descendants of Cain. These intermarriages led to the demise of faith in the Creator God to the point that only Noah and his family were left of the true believers. There are several problems with this explanation.

First, the sons of God are spoken of in other parts of the Bible as angels, such as Job 1:6 and Job 2:1. Psalm 29:1 refers to either men or angels as these sons of the mighty *Elohim*. In either case, because the descendants of Cain were in rebellion to God, they would not be considered sons of God in Noah's time for, like Seth, they were in the image of Adam not God. Psalm 89:6 definitely refers to angels as the context makes clear.

Daniel 3:25 tells of the one seen in the fiery furnace with the three servants of God was, "like a son of the gods."

This was certainly an angel or the Lord incarnate, Himself. In the New Testament, men are not children of God until they are born again. Until then, they are children of wrath.

Second, in verse 4 we are told the union of the sons of God and daughters of men resulted in abnormal offspring called, "men of renown." This title suggests super human qualities that go beyond the normal. This would not be the case if men, descendants of Adam, were reproducing with women through the normal process.

Third, there was no prohibition of Cainites marrying Sethites.

Fourth, the text makes a clear distinction between sons of God, not men, and daughters who are born of men. They are each obviously products of two different domains.

The better Biblical interpretation is to understand the sons of God to be fallen angels for several reasons.

First, there are several instances in scripture where angels take on the form of men. The angels that came to Abraham and then went to destroy Sodom and Gomorrah were almost sodomized. Gideon and Sampson's mother described the angels they saw as men that came to deliver God's message to them. Hebrews 12:2 reminds us to give hospitality to strangers who could possibly be angels without our knowing it.

Second, as we have already seen, "sons of God" refer to angels in the rest of the Bible.

Third, this kind of activity would certainly be unnatural and considered an abomination by God.

Fourth, the offspring of this kind of union would not be normal. The term, "men of renown," indicates super human qualities. They could be the reality behind the mythological stories of heroes who had human mothers and a god for their father.

Hercules would be the best known example of this kind of "men of renown."

Fifth, Jude 6-7 is very clear that there were angels that, "...did not keep their own domain (spiritual abode not physical)," and are now being held by God until the final judgment. That would suggest that God put a stop to that kind of rebellion against His natural order and that is why it no longer takes place. It is also a warning that God will finally judge all unnatural sexual indulgences such as homosexuality and bestiality. One interesting side, this may have been one of the ways Satan used to try to defile the seed of the woman in order to thwart God's plan to destroy him.

In chapter 6:3, the Lord says, "My Spirit shall not always strive with man forever, because he is flesh, nevertheless his days shall be one hundred and twenty years." Some say that this verse means men would no longer live more than one hundred twenty years. As we continue in scripture we see that men lived much longer than one hundred twenty years for a period of time, even after the flood. The better interpretation suggests that God was giving the human race one hundred twenty years to repent. This makes sense in the context of what He said in verse 3 that He would not strive with man forever in his fallen state.

This was to be the time frame for Noah building the ark and preaching the coming judgment, giving man ample time to repent (I Peter 3:20). The one hundred twenty years probably signifies three periods of forty years or possibly three generations.

It was ample time for Noah to build the ark and for mankind to repent.

The word, *Nephilim* introduced in verse 4 refers to giants. It is not totally clear if these giants were the men of renown or human giants such as the giant sons of Anak that Caleb defeated in Canaan or the giant Philistine, Goliath, who had brothers in David's time. It may also refer to giant animals, such as dinosaurs, but in context they seem to be unnatural men. One thing we can be sure of is the fossil record supports the realty of giants before the flood. There have been giant human foot prints found in several parts of the earth as well as giant skeletal remains. The major intent of the verse was to make the point that the presence of these giants or men of renown were abnormal and contributed to the reasons for God's destruction of the earth by the flood.

Genesis 6:5 sums up God's ultimate reason for bringing the flood, "Then the Lord saw that the wickedness of man was great on the earth, and that every intent of the thoughts of his heart was only evil continually." This verse established the fact that actions begin in the heart. The wickedness of man (godless acts) was great because man's heart was totally evil, giving no thought to God or His goodness. Man, left to himself, was totally depraved and irredeemable.

The only alternative God had was to wipe man off the face of the earth. To continue to allow man to reproduce offspring would be irresponsible. Mankind was becoming more and more evil with each new generation born only to inherit Hell.

Destruction was God's only alternative. The fact that Noah was the only one left on the earth from the line of light that was worth saving established this reality. God's coming destruction would be an act of justice and grace.

An act of justice to those who had totally rejected God and an act of grace to those who would not be born for destruction.

Genesis 6:7 is one of the saddest verses in scripture where God said, "I will blot out man whom I have created from the face of the land, from man to animals to creeping things and to birds of the sky; for I am sorry that I have made them." This verse takes us back to chapter 2 and the great expectation of God as He formed man from the dust of the ground. What potential there was then for fellowship and a love relationship. Now God was sorry for making man. It is sad that God so desired to bless man, as He still does today, and yet we turn away from Him in distrust and self reliance just as man did back then. Another sad thing is all the animals on land and many in the seas would also have to perish because of man. But that only makes the point that the creation was made for man who was given dominion over it all. What applied to man had to apply to the rest of the creation.

As finite human beings, it is hard for us to imagine God having similar emotions to our own. How does God change His mind when He is all knowing and unchanging in character and purpose? There are other instances in scripture where God changed His mind. He was going to destroy the Israelites at Mount Sinai and raise up another people from Moses; but Moses interceded for the people and God changed His mind. God was also grieved that He put Saul on the thrown of Israel after Saul's continual poor judgment before God. God replaced him with David.

Thankfully, all that God has determined to do through His Son, Jesus Christ, will never change because Christ was totally obedient. That is why we are secure in Him. God dealing with fallen man, who continues to be a free moral agent, forces God to change His mind when men change their hearts and minds toward God.

And yet, we see in the process of changing His mind that God has emotions and does not do so without hurt in His own heart. The scripture does say that God does not desire for anyone to be lost but that all should be saved. God wants to save everyone but, sadly, not everyone desires to be saved, at least not by the only way God can save them, through Jesus Christ.

Every human being alive today should rejoice over what the scriptures record next, "But Noah found favor in the eyes of the Lord." In spite of all the wickedness in the world, there was at least one man whom God found He could use to start over again with. We all exist today because one man made God feel like mankind might be worth giving another shot at developing a relationship with. Does this mean Noah was perfect? No, it only means that Noah was a man of faith and could be trusted by God to obey Him. In Hebrews 11:7, Noah is third in line after Abel and Enoch as a member of the elite line of faith. Hebrews 11:7 comes right after verse 6 that says, "And without faith it is impossible to please Him, for he who comes to God must believe that He is, and that He is a rewarder of those who seek Him."

Noah is given as the first example of that truth after it is stated. Noah was a great man because he stood alone in his faith when the whole world was in rebellion against God. Evangelical Christianity is under attack today because we are seen as intolerant and a threat to a one world order.

If we do not stand together now, the day will come when a remnant will have to stand alone, just as Noah did. That brings us to the end of Noah's toledoth in verse 9 that says, "These are the records of the generations of Noah." The information in the previous verses from Genesis 5:1b to Genesis 6:8 were stated to be the compilation of Noah who wrote it down and passed it on to his sons.

They would then record the account of the flood and post flood events and pass them on to those after them. God's progressive revelation of His word continued to develop as eye witnesses were inspired to record what they knew to be true. The next recorded event would be the most devastating in history.

It is obviously presented as an eyewitness testimony that makes it clear that this was a worldwide catastrophe.

INTRODUCTION TO THE NEXT TOLEDOTH

Genesis 6:9b-12 are introductory verses to the next toledoth that was recorded by Shem, Ham, Japheth as stated in Chapter 10:1a. These verses summarized the information and gave added information about Noah he would not have written about himself. Verse 9 tells us Noah was, "blameless in his time," and also that he, "walked with God," just as Enoch before him had done. It is by no mistake that Noah was listed as one of the three most righteous men who ever lived along with Job and Daniel. The next four chapters, 6, 7, 8, and 9, are testimonies to his faith and obedience.

The summary of verses 10-12 make it clear that this was a writing that was written separately from the previous toledoth and added later in a compilation by Moses.

It states once again that Noah had three sons, Shem, Ham, and Japheth. It also says the earth was corrupt and full of violence, and that, "...all flesh (mankind) had corrupted their way upon the earth." The impact of combining Noah's account with Shem, Ham, ,and Japheth's was the strong emphasis that was placed in God's word about how totally depraved man had become to the point that only one was left who worshiped God with a whole heart.

In chapter 6:13, God announced His decision to destroy man and everything on the earth. The number 13 seems to be the appropriate number for this verse that spells the unhappy ending for man.

That ending was not about bad luck though, it was about bad choices and living godless lives. The violence that God spoke of had to do with the increased shedding of blood that grew out of Cain's murder of Abel and then Lamech's killing of a young man. The corruption described the rampant immorality that must have been at least equal to that of Sodom and Gomorrah that God would destroy later (Jude 7); and the sins of the Amorites, whom God used the Israelites to destroy as He promised Abraham He would do in Genesis 15:16.

The violence spoke of the shedding of innocent blood as the world had become a dangerous place to live. For those who hold to evolutionistic interpretations of the Bible, it would be wise to consider that evolution is a very violent theory of survival of the fittest, pain, suffering, death and the extinction of some species to make room for more adaptable ones.

If evolution is the process God used to create with, then God would contradict Himself by judging the earth for doing what was supposed to be natural through violence. The children's movie, The Lion King, glorifies the process of evolution by making a lion the king of all the animals as they come to pay homage to him.

The song in the movie, "The Circle of Life," actually worships the evolutionary process of life, death, and new life. But why is the lion seen as king of the animals in Africa? Because lions are at the top of the food chain and the hunters and killers most feared by those they violently prey on. The flood is not justifiable if evolution was a part of God's original intent.

If survival of the fittest through the means of tooth and claw was God's approach to creating new forms of life, then violence and bloodshed would be good not evil. It would be contradictory for God to destroy His creation for doing what they were created to do.

On the other hand, you would expect God to judge mankind if He created all things for life and love, and then man as God's steward, in rebellion, totally corrupted that intent into death and hate. The flood philosophically does not fit evolution theory. It does fit a God of justice who made all things good and that good was diminished into evil by fallen man. That corruption called for justice and a renewal. Remember, God's word must be consistent with itself, not man's distorted understanding of science. The flood and evolution worldviews are not consistent. The flood is consistent with a good God of justice who must punish sin.

FLOOD PREPARATION

God was determined to destroy all life on earth with a flood and begin again with one righteous man, Noah, and at least two of every living kind of birds and land animals. To accomplish this, God had a plan. Noah would build an ark large enough to accommodate his family, all the animals, and enough food to last a full year and one week. God's plan was for Noah to build an ark according to God's specifications. This is important to understand when we realize that only God knew what was coming (Noah had not seen rain before, let alone a flood). God gives Noah the general idea of the ark and then trusts Noah's ingenuity to engineer the building of the ark in a way that would withstand the physical challenges of a catastrophic flood.

This is a testament to Noah's intelligence, creativity, and thoroughness. Noah was the kind of man that, when given a job to do, he could be trusted to do it right.

He was an Ecclesiastes 9:10 kind of man, "Whatever your hand finds to do, verily do it with all your might..." We know Noah did just that because we will see later that the ark survived the rigors of the year long flood.

Genesis 6:14-16 give God's specifications for the ark. It was to be made of gopher wood, a name of a wood not known today, but it must have been dense and hard to hold up to pounding waves. The inside of the ark was to be made up of a menagerie of rooms filling three decks. This would triple the floor space and provide an inward skeletal support system that would support the outside walls from all directions. All the wood was to be covered with a pitch made of some kind of resin or coating that would saturate the wood and make it water resistant much like treated lumber we use today for outdoor building projects. It would also provide a seal between the outside wood slats of the ark.

This word for pitch was the same word translated as covering or atonement, suggesting the ark would be the covering for those in the ark against God's judgment.

This large three story barge was to be 300 cubits long, 50 cubits wide and 30 cubits high. The problem is, we do not know which cubit Noah used from antiquity. It ranged from 24 inches to 17.5 inches. Dr. Henry Morris used the smallest cubit of that time, 17.5 inches, and determined the ark was 438 feet long, 72.9 feet wide, and 43.8 feet high.[36] It may have been much larger but not any smaller than those dimensions. According to these specifications, two major conclusions can be drawn about the ark. First, this kind of floating box can be shown to withstand being capsized, even when forced to a 90 degree angel in highly turbulent water. Second, the inside capacity would equal 522 box cars. That means 240 animals with the average size of sheep could be carried in one box car; 125,000 could be carried in the ark.

Remember, the largest cubit was up to 24 inches and the smallest 17.5 inches. The ark could have been much larger than what is speculated here.

Verse 17 begins with, "And behold, I even I am bringing the flood of water upon the earth..." this is the first mention of the word flood. In the Hebrew it is the word, *Mabbul* and is only used to refer to this historical event. It could be translated catastrophe of water or cataclysm as in the Greek New Testament. In the rest of the Bible when this word is used, such as in Psalm 29:10, "The Lord sat as King at the flood..." it refers to this worldwide catastrophic event. This verse goes on to make it clear that this was to be a worldwide flood that would destroy all living things on the earth when it says, "...to destroy all flesh in which is the breath of life, from under heaven; everything that is on the earth shall perish."

This verse as well as those coming in chapter 7 will make it clear that anyone who tries to say this was a local event must force their own worldview on this narrative rather than just let it say what it says and mean what it means.

Genesis 6:18 is a very important verse of scripture. It reveals one of the most important lessons we can learn about the meaning of the life of Jesus Christ. The verse says, "But I will establish my covenant with you; and you shall enter the ark - you and your sons and your wife, and your son's wives with you." The lesson is this. God only established His covenant with Noah. The covenant referred back to Adam and Eve and God's promise to destroy Satan's curse through the seed of the woman. The seed would now come from Noah who was a descendant of Adam. Only Noah was declared by God to be righteous and blameless in his time, and the reason why was that only Noah walked with God. It was because, "Noah found favor in the eyes of the Lord," that there was going to be an ark and anyone was going to enter that ark.

Out of his righteous obedience to God, Noah saved himself and his family. But here is the point.

Noah's family was on the ark because of Noah's righteousness, not because of their own righteousness. The same is true of Christians. We are not going to survive the wrath of God's coming judgment because of our own righteousness. We will be safe because we are in Christ Jesus by believing in His righteousness, not our own. Notice, this toledoth began with chapter 6:9b that states Noah's righteousness and is written by his sons who understood they were saved by their faith in their father's relationship with God, not their own.

We are saved when we put our faith in the righteousness of Jesus, who was without sin and rose from the grave to prove it. Noah's family was saved by the righteousness of Noah and the grace of God. God's family is saved by the grace of God through the righteousness of Jesus Christ. This is an important New Testament doctrine we find established here in Genesis. God's word is truly consistent with itself.

In Genesis 6:19-22, God instructs Noah to take two of every kind (male and female) of all the animals including birds and creeping things to keep them alive. To keep them alive would suggest all the other living things would die. Those who question this account ask questions like, "How did Noah get all the kinds of animals from around the world such as polar bears or kangaroos to the ark?"

First, in verse 20 it says that all these animals, "Shall come to you." Noah was not going to have to go and round-up all these animals and bring them to the ark, God was going to use the animal's instincts to draw them to the ark. Just as God brought the animals to Adam to name, He would also bring them to Noah to protect in the ark.

Second, because the earth, at that time before the flood, had no barriers to separate the kinds of animals like we have today, all the different kinds of animals were evenly distributed all around the world.

There was no need for adaptation at that time because the environment around the world was fairly much the same. The animals would not have to come from all over the earth because every kind lived in close proximity to where Noah built the ark.

Third, God only required two of every kind, not the different variations within each kind. That means only a male and female dog, a male and female cat, a male and female horse, and so on.

All these kinds would develop variations after the flood as they repopulated the earth. Some would become extinct while others would later be isolated by land and sea barriers, such as marsupials in Australia or pandas in China. The animals coming by kinds would lower the number of animals that would need to come on the ark.

Verse 21 brings up another controversy. It says, "And as for you, take for yourself some of all food which is edible, and gather it to yourself ; and it shall be food for you and for them." Notice the, "And as for you..." phrase. This suggests that as God did the work of bringing the animals to the ark, Noah's family would have to do the work of gathering food for themselves and the animals. That was their task after finishing the ark. What would that food be? We were told in Genesis 1 that man and all animals were to eat vegetation, not meat. We do not see that mandate changed until chapter 9. If the food was to be nothing but vegetation, that would make sense for several reasons.

First, if some animals were meat eaters, Noah's family would have to bring on other animals beside the two by twos to feed the meat eaters.

Second, it would be a task keeping the meat eaters like lions or T-Rex from killing off the other animals. If they were all herbivores, this would not be a problem at all.

Third, all the animals would need to be juveniles at least so they would not take up as much room and would not require nearly as much to eat as adults. This would be especially true of the huge animals such as long necked dinosaurs and elephants. Remember that the average size of the animals was probably the size of a sheep.

They also would not need to be ready to reproduce until after they left the ark.

Fourth, There was also a possibility that some of the animals would hibernate for at least part of the year so they would not need food during their hibernation. This would also lower the amount of food needed to be gathered.

Genesis 6:22 gives the secret to God's choice of Noah as being righteous, "Thus Noah did; according to all that God commanded him, so he did." God knew Noah would obey Him when He gave Noah a job to do. At first glance, you would think that Noah was saved by his works. But as you meditate on it you will realize that Noah was a perfect expression of faith. James 2:17, makes the point, "Even so faith, if it has no works is dead, being by itself." Paul speaks of Abraham as being declared righteous by God because he believed God in Genesis 15:6. Abraham obeyed God in many things to show He believed. In the same way, Noah was righteous because he was a man of faith that showed his faith by his obedience to God.

Thus, Noah had great faith that he expressed by his obedience. Remember once again that Noah is listed third in the faith chapter, Hebrews 11. Noah definitely was saved by faith.

THE FLOOD

Genesis 7 covers the time from the week prior to the beginning of the flood to the waters completely covering the earth for 150 days.

In the week prior to the flood (verses 1-5), Noah and his family needed to know that it was time to load the ark.

With all the food and the number of animals they were dealing with, it would take at least a full week to fill the ark, especially if they honored the seventh day of rest. Their work was increased because, in chapter 7, God added the requirement of taking seven extra pairs of clean animals and birds that would be used for sacrifices after the flood. Verse 1 gives the command to begin to enter the ark seven days before the flood. In this verse it is stated once again that they were all allowed to enter the ark because Noah alone was found to be righteous on the earth. Verse 5, once again, states the reason Noah was declared righteous, "And Noah did according to all that the Lord had commanded him."

Genesis 7:6-10 tell us that Noah was six hundred years old when the flood came. According to the chronology given by Noah in chapter 5, the flood came 1656 years after the creation of Adam. As we have already said, it was the same year that Methuselah, Noah's grandfather, died at the age of 969 years. These verses also reaffirm what was already said in verses 1-5. All the animals entered the ark by twos, there being seven pairs of clean animals and one pair of unclean animals.

Those animals would all be escorted to their particular room God had commanded Noah to prepare. After seven days of loading, verse 10 tells us that the waters of the flood came.

Genesis 7:11 is a key verse. It tells us the very day the flood came and exactly how it all started. The flood came in the 600th year of Noah's life, in the second month, on the seventeenth day of the month. If the account was following the Jewish calendar, which starts with the Passover in early spring (Exodus 12:2), then the flood began in spring.

This is important because we will see later that they will leave the ark a year later in spring when the olive branch began to bloom. That would give them time to prepare for the first winter that was to come after the flood (Genesis 8:22). We have seen precursors to other parts of the law, such as the Sabbath, clean and unclean animals, and a sacrificial system. There may have been distorted calendars that God corrected when He established the Passover month as the first month in the Jewish calendar. Most scholars agree though, that the oldest calendars started in early fall. Suffice it to say this is an account given in a historical context within a specific time frame.

The other important thing to see in verse 11 is the account reported the beginning of the flood in exactly the way you would expect it to be reported on the basis of the Biblical description of the environment of the earth given in Genesis 1. That account tells us that God started with a ball of water and then raised some of that water (vapor canopy) above the atmosphere that surrounded the sphere of water below. It then tells us that the dry land appeared out of the water, trapping water in underground aquifers, while some of the water was gathered into seas. Verse 11 says, "...on the same day the fountains of the great deep burst open, and the floodgates of the sky were opened."

Notice, the pressurized water under ground exploded into the sky forcing mud, rock fragments, and dust into the upper atmosphere that would then cause the water vapor in the canopy to condense and fall in a torrential rain.

The blast of the fountains of the deep would also cause fissures in the crust of the earth that would bring about volcanic explosions. All this expulsion of debris into the water vapor canopy would be necessary for the water there to fall to earth as open floodgates of rain that took forty days and nights to dissipate as verse 12 indicates. The water and steam from within the earth would create rain as well as it fell back to the earth.

But, this all began in such a way that caused a rapid rising of the waters, and yet, not so fast as to endanger those in the ark. God raised the pressure just enough inside the earth to crack its basement rock crust into fissures that would eventually lead to continental drift as the earth's surface began to move under the new pressures forced on it.. The water exploded out high into the canopy and the canopy condensed into rain. That description follows the Biblical picture of the pre-flood earth perfectly. God's word is consistent with itself.

Chapter 7:13-16 once again states precisely all those who where in the ark when God brought the flood. Verse 16 once again reminds us that this was all done as God had commanded Noah. The verse ends with a very sobering statement, "...and the Lord closed it (the door) behind him." Apparently Noah was the last to enter the ark to make sure everyone that God commanded him to bring on the ark got in. That reminds us of what Jesus said that all of those who God has given to Him, He had lost none. It also makes it clear that this was God's judgment and it was He who would shut the door. The closing of the door by God Himself not only tells us that He sealed those inside, but that He shut those out, who were left on the outside.

There will come a time when God shuts the door of grace and those outside of Christ will face judgment.

The flood story is a strong reminder of God's judgment to universalists and pluralists, who try to say there are many ways to God or that God would not send anyone to Hell, because everyone is already saved.

The ark was God's only provision for salvation from the flood and all those outside the ark died.

In the same way, Jesus Christ is God's only plan of salvation and those who do not accept Him as their Savior and Lord will spend eternity in Hell. That is not my opinion; that is what God's word says and illustrates here.

Chapter 7:17-24 are verses committed to convincing their readers that the flood was a worldwide flood, not a local flood. The waters increased and the ark began to rise with the water. It is hard to imagine what those inside the ark heard going on outside the ark before the water got so deep that nothing could be left alive. There must have been cries for help; whaling, shrieks and groans from animals and humans alike. It would have surely been a bumpy ride for a very long period of time. There would have been a lot of adjusting to be made, especially for those who might have gotten seasick. There would be much to be done on the ark to keep everyone busy in the long days and nights to come.

There are several phrases in these verses that make it clear that the flood covered the whole earth. First, verse 18, "...the water prevailed and increased greatly upon the earth." The word "prevailed" suggests an overtaking of the earth. Second, in verse 19 we are told, " the water prevailed more and more upon the earth, so that all the high mountains everywhere under the heavens were covered." The waters totally immersed all the high mountains under heaven.

The most elemental law concerning water is that it seeks its own level. A local flood would not cover all mountains. It would only cover those in the basin it took place in but not those surrounding the basin. The water would have to rise to cover all mountains or no mountains at all. Mount Ararat, located in northern Turkey today, is a high enough mountain (17,000 feet) that for the water to cover it by 15 cubits all flat lands on earth would have to be covered and most mountains.

But the text tells us that all mountains (including Everest) were covered with water, which makes the most sense when we know water seeks its own level. Third, verse 18 tells us the water prevailed 15 cubits higher and covered the mountains. The ark was 30 cubits high. As it floated in the water it must have had 15 cubits submerged in the water and 15 cubits above the water. The point was that the bottom of the ark did not drag or get caught on any mountain tops. Fourth, the word "all" is used four times in verses 21 and 22. All animals died; all mankind died; all that was on dry land died; all in whose nostrils was the breath of life died. In other words, there were no survivors left on the earth outside of the ark.

Verse 23 sums it all up, "Thus He blotted out every living thing that was upon the face of the land, from man to animal to creeping things and to birds of the sky, and they were blotted out from the earth; and only Noah was left, together with those that were with him in the ark." How many times does a fact have to be stated in how many ways before it finally makes its case? The fact of the worldwide flood and the total loss of life, besides those on the ark, is over stated in these verses to the point of redundancy, but done so, obviously, to drive home the point; this was a global catastrophe.

As verse 24 states, "And the water prevailed upon the earth one hundred and fifty days."

After the rain and underground eruptions slowed, the water covered the earth for 150 days before it began to recede. That is a period of five months.

Let us ask some questions to show how nonsensical it is to say this was a local flood.

Before we do, we need to remember the earth was totally populated worldwide. The animals were created in hosts and commanded to fill the earth. On the basis of age length and the size of families already described, Henry Morris calculated that there could have been as many as 7 billion people on the earth before the flood and this was a conservative figure.[37] Remember, the life expectancy before the flood was hundreds of years. Adam lived long enough to see nine generations come after him. Now days, five living generations is very rare.

Families were also much larger then. Genesis 5 tells us that each of the men listed there had several sons and daughters. The pre- flood population growth would have been much faster than it is now. The present earth population is at six billion. It has taken four thousand years from the flood to grow that large. The flood came 1656 years after the creation (almost half of the time from after the flood to the present). With much longer life spans and much larger families, seven billion people is a reasonable estimation.

With this in mind, here are our questions about the foolishness of a local flood explanation of these verses.

First, why build an ark to escape a local flood? Just take a trip far enough way from where the flood was to occur.

Second, why take animals on the ark? There were enough animals on earth to replace those that would have drowned in a local flood.

Third, would not humans have enough sense to escape flood waters that did not engulf them immediately? Some would perish but others would survive as they escaped to the outer perimeter of the local flood waters. The flood eyewitness account tells us every living thing died on the face of the earth. Either a local flood is true or a worldwide flood is true. They both cannot be true. They are opposite concepts. I am going to stick with the eyewitness account.

The ark makes no sense in a local flood story. A local flood brings into question the historical reliability of God's word. If those committed to "old earth" explanations of geology want to reject a worldwide flood, they can do so according to their own worldview, but they step out of bounds when they try to make the scriptures fit their point of view. If the flood is true though, then all geological formations and the fossils in those formations can be explained by that flood with far more reasonable results. The only reason to believe in a local flood is because a global flood threatens "old earth," evolution based worldviews. Honest theology will just let God's word say what it says and then let it guide our understanding of truth, including geology.

FLOOD EVIDENCE

There are many evidences in geology that point to the flood. First, the vast majority of all rock on the surface of the earth is sedimentary rock. That is rock formed by sediments collected in water that then hardens when the water is removed. Most fossils are found in sedimentary rock. Other rocks are formed by lava flows out of volcanos. Much of this volcanic rock was formed under water.

Basement rock that once made up the crust of the earth is now broken into techtonic plates that still are adjusting thousands of years after the flood causing earthquakes all over the earth.

Second, all fossils are formed by plants and animals being buried rapidly in water, sediment, and heat. There are mass burials of animals all over the earth that are best explained by a flood catastrophe.

Third, fossils of animals found at the bottom of oceans like clams and snails, as well as fish of all kinds are found in sediment rock on the tops of mountains.

The mountains had to have been under water at some point in time for those water creature fossils to be there. It is also interesting to note that most geologist believe the mountain ranges were formed fairly recently. How about four thousand years ago, the creation model would ask?

Fourth, the Cambrian explosion at the bottom of the geological column, where all phyla of life forms are found buried together, is explained far more reasonably by rapid burial during a one time event. The Cambrian level does not have any pre-existing life before it and there are no intermediate life forms between those that are found. In other words, there is no evidence of any evolution processes taking place in the fossil record. Why? All fossils in the fossil record were buried at the same time in a year long flood four thousand years ago.

On May 18th, 1980, an event took place that totally and completely turned the concept of uniformitarian geology on its head. On that day, Mount Saint Helens, in the state of Washington, blasted away many assumptions held by old earth geologists at that time. The initial landslide and subsequent steam blast sent tons of earth into Spirit Lake at the base of the mountain.

This caused a tidal wave to cross the lake and hit the side of the mountain on the other side of the lake. In a matter of seconds, a whole forest of trees was washed off the side of the mountain and back into the lake, leaving a huge floating mat of trees.

A vast mud flow spewed out of the crater that swept mercilessly into the valley removing every thing in its path for miles. A cloud of debris burst at a rate of a mile a minute into the open sky that created a cloud cover of silt and ash that covered most of the northwest states for several days. This cloud cover actually lowered the temperatures for months.

Several trips to the active volcano by "young earth" geologists after the first eruption and then others that came later revealed some important observations. First, the initial mud flow (May 18, 1980) was cut through by another mud flow not long after the first (March 19,1982). It left a miniature grand canyon complete with strata layers hundreds of feet deep and a river running through the canyon. But, the canyon caused the river rather than the river taking millions of years to cut out the canyon.

The strata were not formed by millions of years of slow deposits. It was the undulation of the mud flow that caused the strata to form as it slowed and sped up while cutting through whatever got in its path. Second, Dr. Steve Austin, a specialist in the formation of coal was able to use the floating logs in Spirit Lake to show how coal can be formed rapidly as the bark from the trees fell to the bottom of the lake and began to be buried rapidly. This is a much better explanation of coal formation than the slow burial of pete bogs used by "old earth" geologists. Third, as the trees in the lake began to get water logged, they would sink to the bottom by their root ends first. While on the bottom of the lake they would begin to be buried by sediment.

This has become a study in the transference of trees from their original growth spot to another spot and then be buried only to become fossilized. It has become a perfect explanation for the fossil forest standing in Yellowstone Park. It also explains polystrate fossils in strata.

There are many other evidences for the worldwide flood. The fact that many geologist are now becoming committed to local catastrophes makes it clear that the whole earth is covered with geological formations that can only be explained by rapid development due to water and volcanic disruptions.

When you think of the earth being covered with much larger Mount Saint Helens volcanos erupting all over the world at the same time and even continuing under water after the water covered the earth, one can only begin to imagine the total destruction of the pre-flood world. All that destruction left a completely new surface that formed under the force of water and the upheaval of the crust of the earth. It is useless to search for the garden of Eden, the tree of life, the tree of knowledge of good and evil, the four rivers that flowed out of Eden, the city built by Cain and Enoch, and who knows how many other cities that were destroyed and buried miles below the new surface of the earth after being totally obliterated. We are fortunate to find buried relicts here and there like a hammer head, iron pot, or gold chain that can only give a small glimpse of the world that was before judgment came and took it all away.

Recently, The Institute in Creation Research (ICR) completed an eight year study on radio isotope dating. It was mentioned earlier concerning the study of the Hebrew text of Genesis 1-2:3 being a historical narrative rather than poetry.

The major emphasis of the study was to show how radio isotope dating used by old earth geologists to give ages of rocks dating in the millions and billions of years.

The results of their studies can be found in the book, *Thousands...Not Billions*, compiled by Dr. Don De Young.

One of their major discoveries was traces of carbon 14 in diamonds. Because of carbon 14's short half life (5730 years) there should not be any left in carbon containing substances older than 100,000 to 250,000 years. Diamonds are made of coal, pressure, and heat, and are considered some of the oldest substances on earth. And yet, the diamonds studied contained enough carbon 14 to suggest they were only a few thousand years old.

That is what you would expect to find if coal seams were buried during the flood. There was carbon 14 found in coal, oil and dinosaur fossils as well. The findings suggested all these things were buried at the same time and that the amount of carbon 14 in the pre-flood atmosphere was much less than it is now. Another major inference was that radio isotope decay rates were drastically increased during the flood; this undermines one of radio isotope dating's major assumptions that decay rates have been constant for millions of years.

The ICR studies present solid evidence that radio isotope dating methods are unreliable for scientific research. The method's used to give dates of millions and billions years are products of untestable assumptions.[38] Let me remind you that the above findings are predicted by the creation model.

What are those untestable assumptions? Before we answer our question, we need to understand what isotope decay is. There are several elements that can be used in radio isotope dating methods such as uranium to lead, potassium to argon, rubidium to strontium, and so on.

These elements decay by losing alpha or beta particles in their atoms over time that change the elements into what is called their "daughter elements". For example, uranium is the parent element that decays into the daughter element, lead.

Their decay rate is broken down into half lives, which means that during that time the parent element has decayed to where only half of its original amount of element in the rock or fossil remains while the rest has decayed into the daughter element. The more daughter element and the less parent element there is the older the rock or fossil is determined to be.

Now back to our question about untestable assumptions used in dating methods.

First, when a scientist wants to date a rock or fossil, he must assume the rock or fossil has remained in a closed system during its time of decay. That means that it must be assumed that no elements have been leached in or out by various processes of liquids, heat, acids or other forces flowing through the rock. The problem is that there is no way of knowing what all a rock or fossil has been through over thousands of years, let alone millions of years. There is no such thing as a closed system when it comes to any objects buried in the earth.

Second, the scientist has to assume there were no daughter elements present when the rock or fossil was buried. Again, there is no way of knowing what the original amounts of the elements were when the rock was formed or the fossil buried.

Third, we have already made the point that it has to be assumed that the decay rates of the parent elements into the daughter elements were constant from the time the rock was formed until it was tested for age.

There is no way of knowing what would have taken place over millions of years that would effect decay rates.[39]

The ICR studies discovered the decay rates may have been much faster at the time of the flood, showing decay rates are not constant.

Fourth, these assumptions have to assume that uniformitarian geology is true, which is something that cannot be observed in the past nor confirmed in the present, and thus is not verifiable.

Fifth, scientists have to assume the worldwide flood did not take place in the face of the fact that there is far more evidence to support it having taken place rather than not. Sixth, they also have to assume that there was no water vapor canopy before the flood, which would have had a measurable influence on element deposits and their decay rates.

Finally, they also have to assume the creation week did not take place as recorded in Genesis 1, which, if true, would bring into question what the element ratios (parent elements to daughter elements) were to each other as they were all created together. All these assumptions make the point that radio isotope dating is more about faith than science. They have to believe their assumptions are true because they cannot prove, by scientific observation, that they are true.

Here is a case about how bad assumptions can lead to bad science. When the dating method of carbon 14 was being developed back in the 1950's, it was assumed that carbon 14 was in equilibrium in the atmosphere. That means there was as much carbon 14 coming into the atmosphere as was being lost. It had to be established that carbon 14 was a constant in the atmosphere that was being absorbed at a constant level by plants and animals in order to get good dates. It was determined by scientists that it would take 31,000 years for carbon 14 to reach equilibrium in the atmosphere.

These scientists, being biased by the "old earth" assumptions that the earth was 4.5 billion years old and that uniformitarian (the present is the key to the past) dictum was true, assumed carbon 14 had been in equilibrium for ages.

The fact of the matter is, carbon 14 was discovered to not even be close to equilibrium. Its levels actually suggest the atmosphere is still quite young in its carbon 14 development. Bad assumptions lead to bad conclusions and that is bad science. Carbon 14 dating is not nearly as reliable as evolution scientists would have us believe because it is based on faulty assumptions.

There were two other important discoveries by the ICR studies. They were the discrepancy of dates in rocks and the amount of helium left in zircon crystals already mentioned in chapter 5. The ICR team took some rocks and dated each rock using several dating methods on each rock. If the methods were valid, they all should have given the same date for the rock being dated.

The fact is, each rock was given dates that varied by millions of years according to the different dating methods. If these dating methods were reliable, they should have all given the same dates.[40] They also found that helium amounts in zircon crystals taken out of granite rocks showed the crystals could have been in existence for only about 6,000 years.

Helium leaks out of whatever substance it is caught in like the helium in party balloons. If the crystals were millions of years old, the helium would have worked its way out of them a long time ago. The fact is, there were large amounts of helium still trapped in the zircons.[41]

Dr. Henry M. Morris, in his small but powerful book, *The Scientific Case for Creation*, gives a list of 70 ways to measure the age of the earth on the basis of uniformitarian estimates using the three major dating method assumptions (closed system, no daughter elements, constant decay rate). The oldest date given is 260,000,000 years.

Most dates are only in the thousands of years. The point he makes is, even when you use different means of dating besides radiometric dating, and follow the old-earth assumptions to get your dates with, they still give "young earth" dates not even close to one billion years let alone 4.5 billion years, the assumed age of the earth. In reality, no dating methods can be made to be totally reliable, but the fact is, most dating methods support a "young earth" far more realistically than an "old earth."[42] This is another creation model prediction.

CHAPTER 14
WHICH NEW WORLD?

The flood waters continued for 150 days before they began to subside. The last chapter took us from the preparation of the ark to the end of the five months the ark floated in the waters above all the mountains. Genesis 8 will now take us through the dissipation of the water to the time of the coming out of the ark. We will see that the ark floated on top of the waters for five months before coming to rest on top of Mount Ararat. The fact that they were in the ark a total of one year and a week tells us they stayed on the ark another seven months and a week before leaving the ark. But by the grace of God, at least during these seven months, the ark was stable and no longer being tossed about in the water. It also tells us that it took much longer for the water to subside and the land to dry than it did for the water to cover the earth. It must have been a trying time waiting inside the ark until God finally gave the command to leave the ark.

THE ARK RESTS ON ARARAT

The first recorded statement of Genesis 8:1 says, "And God remembered Noah..." There is a contrast here. He remembered those on the ark but not those who died in the flood. Actually this was a remembrance for deliverance. God remembers all those who died in the flood as we are told in I Peter 3:19-20 and Jude 6; but that remembrance is for final judgment.

God remembering those in the ark can be compared to those who's names are written in the book of life, in Revelation 20:15. God will remember all those who, through faith in Jesus Christ, will be delivered from the eternal lake of fire. After Jesus fed the five thousand, He sent the disciples away to cross the sea of Galilee while He stayed there alone to pray. A terrible storm came and the disciples feared for their lives. But in the middle of the storm Jesus came walking on the water and delivered them from the storm. God knows when He is sending us into a storm and He can be trusted to remember to deliver us, in His time, from the flood, the storm, and especially the final judgment. If we are faithful to get in the ark, God will be faithful to make sure we get out of the ark safely. Jesus Christ is our ark of safety Who will carry us through the final judgment of God.

Genesis 8:1-5 summarizes what took place in order for the water to decrease on the earth. Verse 1 says that God sent a wind to evaporate some of the water. This is the first mention of wind in scripture. You would expect this to be the case, especially at that point in time, when the canopy was removed and volcanic clouds of ash still covered the skies. Cold air from the poles of the earth would begin to move against the warm air coming off of the waters that were still very warm from the pre-flood water, the waters from the steam blasts when the fountains of the great deep broke open, and the volcanic magma and lava flows that continued under the water. It must have been a terrifying experience to hear the winds of a storm for the first time. Fortunately they already had the promise of God that He would bring them safely through the flood.

Verse 2 reminds us that the fountains of the great deep and the flood gates of the sky were closed.

We already have been told that it rained for forty days and nights; so that would mean they floated on the water for 110 days after the rain stopped. The water covering the earth would limit the amount of steam and lava blasts coming from out of the ground that would no longer be sent high into the atmosphere before the flood waters came. Verse 3 restates that the water began to recede after the 150 days of the water prevailing on the earth. Verse 4 then gives us the exact time the ark came to rest on Ararat, in the seventh month on the seventeenth day. In Genesis 7:11, we were told the flood began in the second month on the seventeenth day. That would be exactly five months that the ark was in the water before coming to rest on Ararat. That was three months longer than it took the Pilgrims to sail to America on the Mayflower.

Ararat was a very high mountain, and so, even after the ark settled there, it would take another seven months for the water to clear and the land to dry. Verse 5 tells us that on the tenth month on the first day of the month the tops of other not so high mountains could be seen. That statement is consistent with the topography of the land around Ararat. It is one of the higher peaks in that area. It is interesting to note the name, "Ararat", continues to this day just as it was in the days after the flood. This fact is a strong support for the authenticity of the account.

We have to ask ourselves, "Where did all the water go?" The answer to that question is found in Psalm 104:8-9. Speaking of the flood waters these verses say, "The mountains rose; the valleys sank down to the place which Thou didst establish for them. Thou didst set a boundary that they may not pass over; that they may not cover the earth."

These verses give us a picture of the process taking place under the water while the ark floated above. Sediments began to build up all over the earth's surface as large caverns of water from the fountains of the deep were emptied under the crust and began to cave in from the weight of the water and sediment that pushed down on the weakened earth's curst. The caverns would have been created by the emptying of the fountains of the great deep, and other shifting of the earth's crust. These verses actually give the picture of God pressing down parts of the earth's crust that then caused other parts to rise up. It would have been much like needing bread dough. As you push it down in one part, the dough around your hands pushes up.

In other words, the continents began to rise as the ocean floors gave in to the pressures of the added weight of the water and sediments. After it was all said and done, the surface of the earth was covered with two-thirds water and only one-third land. Rather than all lands being connected, the continents were separated by large bodies of water; the deepest point in the Pacific Ocean being seven miles deep (Mariana Trench) and the highest mountains (Everest in the Himalayas) being five miles high. Massive burial sights of land and sea animals began to decay into oil deposits while others began to fossilize. Vast areas of vegetation, such as whole forests floated in large mats while others were buried in sediment during the continual upheaval of the flood waters mixed with mud flows, volcanic lava flows, and mega-tons of debris tossed to and fro in the turbulent waters. Nothing was the same.

In Genesis 8:6, we are told Noah waited another 40 days before opening the window of the ark to send a raven out to find dry land. The raven being a solitary bird did not return but we know it survived.

The raven was an unclean bird so there would only be two on the ark. There are still ravens today so the one let loose must have found places to rest before being reunited with its mate. Noah also sent out a dove, a non-solitary bird that returned because the land was not yet dry enough. Seven days later Noah sent the bird out again and this time it came back in the evening with an olive leaf in its beak. This was a hint that the land was beginning to dry and sprout some vegetation. When the dove was sent out again in verse 12 after waiting seven more days it did not return, meaning the dove found a place to begin nesting rather than returning to the ark. It waited for its partner to come to its new nest. Since that time ravens have become identified with evil and doves identified with good. The raven caring only for itself and the dove returning to those in the ark.

What is the meaning of the sending out of the birds? The most practical explanation would be that Noah was not going to leave the ark until God told him to, but the birds helped Noah and his family to know that the time of their departure from the ark would be soon. It is a picture of hope and anticipation suggesting there is nothing wrong with knowing the signs of the times when we are waiting on God. Finally, Noah was ready to remove the covering of the window of the ark (verse13) to where he could get a good look out. He saw the ground was dried up as far as his eyes could see.

Then, in the second month on the twenty-seventh day, the earth was dry. If the months were 30 days long and they got on the ark in the second month on the seventeenth day and left the ark on the twenty-seventh day of the second month that would be 371 days or 53 weeks; a total of one year and one week spent on the ark.

The important point to be made is found in chapter 8:15 and 16. These verses tell us that Noah was unwilling to leave the ark until God told him to leave. Noah got on the ark when he was told to and he did not leave the ark until he was told to. This is another sign of Noah's pure obedience and surrender to God's will. He may have been under pressure by his family and the stirring of the animals to get out of that cramped up ark sooner, but Noah waited on God. The lesson is, you stay out of trouble and avoid a lot of frustration and heartache when you do not get ahead of God by giving in to outside pressures.

THE EXODUS

Genesis 8:17-19 records the exodus from the ark. As they left the ark, God's command to be fruitful and multiply filled their hearts with hope for the future. Every animal, whether land or bird, went safely out of the ark in the same twos as they had entered in. Verse 19 says they, "...went out by their families from the ark." Noah and his family left the ark as well. They were all in one piece, none the worse for wear. It had been a long journey with many challenging times. But all were safe; none had been lost who were in the ark, just as God had promised. The same is true of those who are in Christ Jesus.

He will bring all those who trust in Him safely through the final fiery judgment still to come, just as I Peter 3:18 tells us, "For Christ also died for sins once for all, the just for the unjust, in order that He might bring us to God..." What a trip that must have been coming down off of that high mountain. They walked on land that was mainly made of sediment layers of rock that was full of former living things now buried and fossilized under their feet.

There were clouds in the sky and the wind was probably blowing. The further they descended, the more barren the land must have looked. The land around Ararat is still an arid desert today. It was nothing like the earth was before the flood had come. No forest of trees or lush rolling plains; only patches of grass and small sprouts of trees and bushes barely coming out of the ground. That would have been one reason why God waited so long to let them come out of the ark. There needed to be enough vegetation to have begun to grow to provide food for them all. They would need that food in the days ahead as they made their transition of adapting to the new environment they were faced with.

The first thing Noah did when he came off of the ark was to build an altar to give thanks to God. This is another one of those indicators of Noah's strong faith in God. He realized the severity of the whole flood event and recognized God's great gift of grace to him and his family by allowing them to survive. And so, the most grateful thing to do was to offer a sacrifice of thanksgiving to God by using some of the clean animals who came on the ark in seven pairs. How often does God grant us a great deliverance and we fail to remember to give Him thanks by giving an offering of ourselves.

We should give God thanks everyday for the deliverance He has provided for us through His Son, Jesus Christ. What Christ did for us on the cross was a far greater act of love then what God did for Noah and his family when He spared them from the flood. Theirs was a temporary deliverance from death but they all had to die in time. Our deliverance from sin and death is an eternal deliverance into a whole new heaven and earth.

Not like the deformed earth Noah and his family faced that was far less than what it had been before God changed it all by the forces of the flood. Chapter 8:21 tells us that God smelled the soothing aroma of the burnt offering and gave the promise to not curse the ground again because of man. The reason was He understood the total depravity of man. His promise to never destroy every living being again seems to indicate that God would now work to bring salvation to all those who would believe in Him.

There is an important principle to remind ourselves of at this point in the post-flood account. I Samuel 15:22 states the principle when Samuel rebuked king Saul for disobeying God by not utterly destroying the Amalekites and all their possessions after defeating them in battle. Saul used the excuse that he was going to sacrifice all the sheep, ox, fatlings, and lambs to the Lord rather than just kill them. Samuel responded with, "Behold, to obey is better than sacrifice." This principle was already established in the life of Noah. It was because of Noah's obedience to God that he was even able to offer a sacrifice to God. Noah's sacrifice had meaning and acceptance from God because he had already obeyed God. Abel's sacrifice was accepted because it was brought out of obedience to God.

Our worship of God is much like bringing sacrifices to God. We often refer to our praise in worship services as a "sacrifice" of praise. The point is this, if we have not been living our personal daily lives in obedience to God's commands, our sacrifice of worship offered to God gives Him very little, if any, pleasure. Noah's sacrifice to God gave God pleasure because Noah's life of obedience already gave God pleasure. God does not want us to just be sold out to Him in church worship services.

He wants us to be sold out to Him in every area of our lives, in every day of our lives. That is the kind of sacrifice God wants from us all. There are signs above our auditorium doors for our church members to read as they leave after our worship services. The signs say, "Let the worship begin." These signs are a reminder that acceptable worship to God takes place in our daily lives as we obey Him each moment in our daily lives. This kind of obedience is what gives our worship in the pew integrity.

Genesis 8:22 is a very important verse to consider. It is a poem recited by God, " While the earth remains, Seed time and harvest, And cold and heat, And summer and winter, And day and night shall not cease." Think of the context. Noah has just come off of the ark after God has just destroyed the earth with water. All was going to be different after the flood. With the canopy gone, there would be a change in weather patterns. Instead of year round even temperatures, there was going to be the seasons of spring, summer, fall, and winter. There was going to be a time to plant and a time to harvest, rather than the year round growing seasons of the pre-flood world.

This poem was a preparation given by God to Noah to get him ready for the changes that were now in place after the removal of the water vapor canopy. This is another one of those confirmations in scripture of its own inward consistency with its own worldview. The poem also reminds us all that with the changes in the weather patterns there would come a shortening of life. Before the flood, people lived hundreds of years.

With the shorter life spans, life would be understood to follow the pattern of the four seasons. Our childhood represents the spring of life when all is new and fresh.

The summer of life is our young adult years where we are strong and full of life. The fall of life is our middle aged years when we start to slow down a bit and our physical bodies begin to show their age like the changing leaves on the trees. And finally, the winter of life comes when everything dies and is buried in the snow. The wonderful thing is that spring comes after winter. Even in the death of winter there is hope of the resurrected life to come. All those who die in Christ can look forward to resurrected life. In this poem of warning of change there is the anticipation of newness of life. Each new year is full of the hope that this year could be the year God fulfills His ultimate purpose to bring in the new creation with resurrected life.

ANOTHER CHANGE

Genesis 9 continues the theme of the new world after the flood. In verse 1, God once again gives the blessing to Noah and his sons, "Be fruitful and multiply, and fill the earth." We have already made this point but it is important to restate it here. It seems that God's vision for man was to remain strongly oriented to family and to live simple lives.

He desired families to grow dependent on eacg other as they spread out over all the land. He did not want them grouping up together and developing large cities. We see this to be in contrast with Cain and Nimrod, a man we will meet later. Both men built cities rather than living simple lives off of the land keeping flocks and herds of animals. The contrast is with Abraham, who was called out of Ur with his father, Terah, and then called out of Haran to live as a sheepherder wondering around the land of Canaan.

It seems that when mankind collected together in cities it increased their capacity for sin. We see a similar trend in the United States that has become less committed to God as more and more people have left the farms and moved to the cities. After the flood, God wanted mankind to spread out in family groups all over the earth but, as we will see later, they decided to fill a city rather than the earth, totally against what God commanded.

Genesis 9:2-3 introduces another change that equals the one given at the fall when God transformed some plants into thorns and thistles. There, the plant kingdom was changed in relationship with man; here, the animal kingdom was changed in their relationship with man. Before the flood, the animals were still in harmony with man and available for his use. After the flood, God changed that relationship. This verse says, "And the fear of you and the terror of you shall be on every beast of the earth and on every bird of the sky; with everything that creeps on the ground, and all the fish of the sea, into your hand they are given. Every moving thing that is alive shall be food for you; I give all to you, as I gave the green plant."

Why did God change the animal's relationship with man from harmony to fear and terror? First, God wanted to protect man from the animals that were going to become meat eating animals and poisonously aggressive animals, such as snakes or insects. Apparently, Noah and his family had nothing to fear when God brought the animals to them on the ark because they were still in harmony with the animals. After their departure from the ark, that would no longer be the case. This change of some of the animals in form and function is not unprecedented in creation.

God continues to change maggots into flies and caterpillars into butterflies. Second, the whole environment of the earth was no longer the same after the flood. Where there were once forests and abundant grass lands everywhere, now there would be deserts and other places covered with ice in the ice age that was to come. Food was no longer going to be abundant year round. In this new life cycle, some animals would eat plants for protein and others would eat meat for protein because of food shortages. There must have been a period of time before this tooth and claw change took place.

All the animal kinds would need some time to reproduce until each kind was abundant enough to withstand being harvested without the threat of extinction by the newly transformed carnivores. Third, man himself would now become a hunter as well as a farmer and keeper of herds. As man moved into regions where the weather was severe, such as the Eskimos in North America or the Neanderthals in Europe, during the ice age, meat at times would be the only food source available to them. Fourth, the fear of man in the animals would also make man have to work for his food just as the thorns and thistles made it harder for man to grow his food before the flood.

It is interesting to note at this point that all animals were to be food for man to eat. There were no restrictions of any kind. The nation of Israel was given those restrictions as a part of their becoming a peculiar people amongst the Gentiles. We find those restrictions were removed in two ways in the New Testament.

First, Jesus, in Mark 7:19, declared all foods clean by making the point that it is not what a man eats that defiles a man but the evil that comes out of his heart.

Second, God told Peter, in Acts 10:15, in the vision of the unholy and unclean animals in the sheet, "What God has cleansed, no longer consider unholy."

This was a definite declaration that Jesus' sacrifice made all things clean and gave the Gentiles as well as the Jews salvation. This was a clear indication that all ceremonial law was made complete in Christ and there is no longer any difference between Jew and Gentile, as Paul states in Galatians 3:28. Things have been restored, by Christ, to this post flood principle of all things acceptable for food. Remember, at this point in time clean and unclean animals only applied to those used for sacrifice, not for eating. The eating restrictions would be given to Israel at Sinai in the Law of Moses. Once again, God's word is consistent with itself.

The one restriction given was no eating of blood. The reason being that life is in the blood. Even though the eating of meat was now to be allowed, there remained the reminder that life was still to be respected. God never intended for death to be a part of His original creation. The life in the blood was to be separated in order to remind man that God is a God of life and life was still to be held in high esteem.

God even gave a warning to show the special kind of life that man holds by saying in verse 5, "And surely I will require your life blood; from every beast I will require it. And from every man, from every man's brother I will require the life of man." Murder and cannibalism was not to be tolerated. By referring to, "every man's brother", it implies that what was the case with Cain or even Lamech and his descendants would no longer be tolerated. A person who took another person's life by murder would have to give up his own life.

Genesis 9:6-7 makes a decree in the form of a poem to strengthen what was already written in verse 5, "Whoever sheds man's blood, By man his blood shall be shed, For in the image of God He made man.

And as for you, be fruitful and multiply; populate the earth abundantly and multiply in it." What this verse did was establish government to protect man from man's depravity. This passage gives government the authority to exercise capital punishment on those guilty of shedding innocent blood. Paul restates this authority of government in Romans 13:4, "...it (government) is a minister of God to you for good. But if you do what is evil, be afraid; for it does not bear the sword for nothing; for it is a minister of God, an avenger who brings wrath upon the one who practices evil." Murder is the most evil crime one can commit. In the Law of Moses, the death penalty was extended to punish other crimes against man other than just murder. Unlike liberal ideology of today, God requires just punishment for crimes committed against society. Jesus forgave the thief hanging on the cross next to His when he asked for forgiveness, but Jesus did not deliver him from the consequences of his sins that put this man on the cross.

It must be understood here that this authority for capital punishment was not given as a deterrent to murder. It was given to show the importance of humanity upholding the sanctity of human life that was created in the image of God. First, people, who do not respect the sanctity of life, no longer respect God and are a danger to society. If they are allowed to live and pass that corrupt attitude on to others, their society will become a violent place in which to live, just as it was before the flood.

Second, any society that no longer respects the sanctity of life will soon destroy itself.

When personal rights and individual choice become more important than human life in a society, that society will soon become subject to the judgment of God.

Before long, that society will be aborting its children; using euthanasia to kill the sick, disabled, old, and finally anyone determined to be a drain on society or no longer able to live a productive life for that society. Without the understanding that a human being is made in the image of God, humans are relegated to animal status and the laws of survival of the fittest kick in. Whoever has the strength to accumulate power rules and then determines who lives and who dies, replacing God with themselves. We only have to look back as far as Saddam Hussein, Stalin, and Hitler (all evolutionists) to see that point made. These men slaughtered millions by claiming the right to do so because they held the power. Respect for life shows respect for God, the giver of life. Those who do not respect life do not respect God and are a threat to the future order of a society.

Once again, God commands Noah and his family to repopulate the earth abundantly. Why was this so important to God?

If we think about it, we must remember where God's word is taking us. At the end of time, God looks forward to having a large family, referred to as the kingdom of God, made up of all those who believed in Him and His redemptive plan finalized in Jesus Christ. These commands to populate the earth, once again, looked forward to you and me and all those who came before us and all who are to come after us that will make up that final kingdom.

But that kingdom will only be made up of those who were born here on earth and then chose to live their lives by faith in God.

It is all about the coming kingdom and all those God looks forward to spending eternity with in His eternal family. He is a big God and has plenty of room for all who will come to Him through His son, Jesus Christ.

THE RAINBOW

Genesis 9:8-17 introduces another addition to the post flood world. Noah and his family, as well as all the animals, went through a very traumatic time while on the ark. They were introduced to torrential rain, massive winds, and thunder storms while the ark tossed to and fro in the open flood waters. Think of the terrible fear those new black rain clouds with thunder cracking and lightening flashing would bring to the hearts of those who survived the flood. They would think that possibly God was going to destroy the earth again. But, God, in His grace and wisdom, gave a sign to all the flood survivors to remind them that He would never judge the earth again with a flood. The sign was the beautiful rainbow that shines in the clouds after the rain. Verse 15 quotes God as saying, "...and I will remember My covenant which is between Me and you..."

This verse takes us forward to Revelation 4:3 where we are given a picture of the throne of God that has a rainbow around the throne. The rainbow in that verse is a reminder of this covenant. The rainbow then, symbolizes for us the sovereignty of God who has complete authority over His creation and reminds us of the faithfulness of God to remember and keep His covenants.

Once again, we are given another evidence of the consistency of God's word with itself. We were told in Genesis 2:5 that God had not sent rain on the earth. No rain would mean no clouds.

No clouds would mean no rainbows before the flood. Some try to say that rainbows always existed but God gave the rainbow this new meaning after the flood. The context of the scriptures referring to the rainbow make it clear that this was a new phenomenon. If it was nothing new, it would not have any significance in the hearts and minds of Noah and his family. This new formation of clouds after the flood also agrees with the concept of the water vapor canopy that would have resisted the formation of clouds. We have already been told in Genesis 2:6 that the earth was watered by a mist rising from the ground and rivers that flowed from out of the ground, like the one in the Garden of Eden. These mechanisms for watering the earth make it clear that there was no need for rain. The rainbow fits perfectly into the Biblical worldview that everything changed after the flood and affirms its consistency with its own worldview.

Genesis 9:18 and 19 remind us that Noah's three sons came off of the ark with Noah and their wives. Verse 18 makes an unusual insertion that Ham was the father of Canaan. Verse 19 then established the fact that all of mankind came from one of these three sons. There were no other hominids out there that evolved into other inferior races of homosapiens as Darwin's evolution would have us believe in his book, *The Descent of Man*. There is only one race of humans and they all came from Noah who came from Adam.

We will see why the subtle addition of Canaan was added to the list of the three brothers later, but suffice it to say at this time that the Holy Spirit, Who inspired the writers of God's history, wanted to make it clear that all mankind came from Adam and then Noah. The fact that all humans descended from Noah's sons strongly undermines the local flood interpretation, mainly because it leaves the door open for other humans to survive the flood in other parts of the world and brings into question the reliability of scripture.

THE CURSE OF CANAAN

Chapter 9:20 and 27 record an event that took place some time after the departure from the ark. We know it had to have been long enough, first, for Noah to settle into farming. Second, there needed to be time for him to have planted a vineyard that was mature enough to produce grapes that could be used to make wine. It usually takes several years for grape vines to mature enough to produce grapes good enough to make wine. Third, Ham's son, Canaan, had to have been born in order for Noah to mention him in his curse and Genesis 10:6 lists Canaan as Ham's fourth son, not his first. It is interesting to think about all the things that could have been recorded in Noah's life after the flood, and yet, this event was the only one God wanted in His word.

What was the event; and why was it included in Noah's life story? The event was Noah getting drunk on his own wine in his tent. He uncovered himself while in a drunken stupor. For some unknown reason, Ham saw the nakedness of his father inside his father's tent and went outside and told his two brothers, Shem and Japheth.

In response to Ham, the two brothers, unwilling to look on their father's nakedness, backed their way into their father's tent with a garment draped over their shoulders and covered their father's nakedness.

Later, Noah woke up from his drunken stupor, but remembered what Ham had done to him. In response, Noah pronounced a curse on Canaan (not Ham) and then, blessed Shem and Japheth. Just as the story of Cain made the point that the fall was passed on to Adam's descendants, this event was a reminder that all of mankind was still fallen, including Noah.

This event being included in God's word is so consistent with the rest of God's word. No matter how godly a man will be portrayed in scripture, there will always be a hint of human frailty to show the fact that man is still a sinner. We see this in the life of the patriarchs, of Moses, King David, the apostles, and many others. While God is accomplishing His redemptive work, He must use fallen men to bring about His purposes. That should be a great relief to us all as we realize that God can use us as well, in spite of all our imperfections. Remember, Noah was a man of faith in spite of his imperfections. God can always use those who have faith in Him.

We do not know what Ham did to his father that caused Noah to remember it, even when he was drunk. Some speculate that Ham preformed some kind of immoral act on his father. Regardless of what took place in Noah's tent by Ham, the fundamental issue was Ham showed no respect or reverence for his father in contrast with his brothers. In our day in time, nakedness is flaunted on every hand. This story makes it clear that modesty and the privacy of one's own body must be respected and protected.

In the temple, the priests wore undergarments that covered even their ankles so that none of their nakedness would show when they walked up and down stairs (Exodus 48:22). Remember our nakedness is the reminder of our fall from purity into sin's depravity.

Ham's act suggests a character flaw in Ham. His lack of respect revealed a possible deeper resentment for his father and his faithful obedience to God. It could be that Ham was feeling the urges of his own depravity to break away from his father's righteous influence. This revelation of his father's human weakness may have given him enough courage to expose his true feelings after living so long suppressing them.

God wanting to reveal Noah's humanity was one definite reason why this story was included here. The other reason has to do with Canaan. Noah's curse of Canaan was prophetic. He was certainly a prophet much like his great grandfather Enoch before him. By building the ark and preaching for 120 years, Noah was prophesying God's coming judgment. Noah does the same here. In Genesis 15:16, God gave Abraham the sign that His promise to make him a great nation would be fulfilled by telling him his family would become slaves for 400 years. After the 400 years, his descendants would then be brought out of that slavery to return to the land of Canaan.

They would then be used by God to bring judgment against all the godless Amorites who were also known as Canaanites. This prophecy by Noah in the form of a curse would be used later by the Israelites as a justification for their destruction of the godless people who were descendants of Canaan.

The descendants of Canaan who moved to the land of Canaan became people who practiced human sacrifice; offered their babies alive on the burning metal altar of Molek; who also practiced prostitution and homosexuality in their temples. It seems that the unholy urges in Ham became more liberally expressed by Canaan, who then passed his rebellion on to his descendants after him.

It is important to make a point here. Only the descendants of Canaan were cursed by Noah. Ham had three other sons, Cush, Mizraim, and Put. They were not cursed. They are not included in Shem and Japheth's blessings, but they were not cursed either. Some became great nations, such as Egypt, Ethiopia, China, Japan, and the many nations of people who migrated to the Americas.

In contrast to all this, Shem was blessed by Noah saying, "Blessed be the Lord, the God of Shem; and let Canaan be his servant." It would be the descendants of Shem who continued to uphold the truth concerning the one true God. The three major monotheistic world religions that uphold the creation, fall, flood, and tower of Babel stories of Genesis are, Judaism, Christianity, and Islam. All three descended from Shem. Canaan, being symbolic of all those who reject God and give themselves to live in rebellion against God, will always fall servant before that which is good and comes from God. Good always overcomes evil in the end.

Japheth was similarly blessed as one who would dwell in the tents of Shem as is stated in verse 27, "May God enlarge Japheth, and let him dwell in the tents of Shem; and let Canaan be his servant."

Japheth became those nations that moved northwest from Mesopotamia and ultimately became the Gentile bearers of Christianity, which came out of Judaism. In other words, Japheth's descendants would come under the protective covering of Shem's faith. Again, Canaan would also be his servant because the ways of God always overcome the ways of man's depravity.

Verses 28-29 finalize the Noah story and the flood. These verses record, " And Noah lived three hundred and fifty years after the flood. So all the days of Noah were nine hundred and fifty years and he died." Unlike the others listed in the chronology of Genesis 5, Noah does not have any other sons or daughters. This is consistent with what has already been stated that the whole earth was repopulated by Shem, Ham, and Japheth. The long life of Noah tells us that having lived in the environment before the flood gave him physical prowess that served him well even after the flood.

As we look into the next chapter, we will see that this longevity continued for sometime after the flood in the lives of those who descended from Shem. If this was also true of animal life, then both man and animals would replenish the earth more rapidly then what present life spans allow. The most obvious deduction that can be made about Noah's longevity is God allowed him a long life, equal to his forefathers. It was a sign of God's blessing on him. It also insured that his testimony of the flood would be established for several generations.

That would explain why hundreds of flood stories still remain in many cultures around the world to this day. Noah and his sons would have repeated it often and passed the story on to further generations just as it is written in Genesis. The story would later be distorted by new generations but still testify to the floods historical validity.

Genesis 10:1 finishes out this third toledoth with, " Now these are the records of the generations of Shem, Ham, and Japheth, the sons of Noah, and sons were born to them after the flood." The previous information from Genesis 6:9 up to chapter 10 would all have been witnessed by Noah's sons and they each would have had input into recording the flood story and their father's faithfulness through it all.

CHAPTER 15
WHICH DISPERSION?

The final two chapters of our study give us insight into where all the nations of the earth originated and why they were dispersed. These chapters stand in stark contrast to "old earth" human history. This evolution based history teaches humans evolved from different pre-man hominids into the different races with slow development of technology like tools, domestication of animals, agriculture, and writing. Studies of DNA and genetic historical development tracing back to our ancestors have shown that all human genes originated from a single source.

The idea that language developed from grunts to words to phrases to sentences to Shakespeare is not supported by the study of languages. We find sophisticated language all the way back to earliest history. Some languages of so-called, primitive tribes of today are more complex than some contemporary languages. Writing, as well, has been shown to be a part of the oldest civilizations as far back as the Sumerians, the oldest known culture. We have already seen that technology was developed just six generations after Adam.

Language, writing, domestication of animals, tent making, agriculture, metallurgy, music instruments, astronomy, city building, engineering of large boats, all existed before the flood and, by all means, were carried over after the flood by Noah and his sons.

Man was highly intelligent (made in the image of God) and that intelligence continued after the flood being found in the dispersion story as well as archeological finds that solidified the reality of that intelligence.

EARLY POST-FLOOD ENVIRONMENT

The world that Noah and his family entered back into after coming out of the ark was very different from what they had known before the flood. The loss of the canopy would cause rapid cooling at the poles, while the oceans remained much warmer than the present. This would cause greater amounts of evaporation and higher levels of rain fall resulting in much heavier erosion than today. Lakes left after the flood, such as in the flat lands of the U.S. plains and in the upper northwest, would fill up rapidly causing breaching of the temporary dams holding these lakes in place. The collapse of these dams would generate the movement of large amounts of water cutting through what ever got in its way as it moved to the seas, leaving behind huge canyons and newly cut mountains, mesas, and plateaus. The Grand Canyon in Northern Arizona would be one of many examples.

Another result of the heavy rain fall in the earth's cold northern and southern regions would be rain turning to snow. Over several hundred years, cooler and cooler summers began to prevail in these northern regions, the snow would not all melt away, developing thick ice glaciers over time. These glaciers would stand for a period of several hundred years before melting back to where they are presently. The present day melting rates make it clear that glaciers melt rapidly not slowly.

Yes, there was an ice age, but only one, and it developed not long after the flood. All the animals we find in the frozen tundra of Canada, Siberia, the Ukraine in their northern regions would have flourished in these previously much warmer areas as warm air from the oceans held back the cold. But then, the oceans began to cool, volcanic ash in the skies held back the warmth of the sun, and shorter cool summers were finally over taken by the winters, as the cold seasons got longer and less snow melted away. We know this glacier activity dipped down into the northern states of America and those in Europe and Asia as well. Michael Oard, in his book, *Frozen in Time*, gives a very convincing explanation of the post-flood ice age in these regions and how to explain frozen mammoths and frozen fruit trees that still have their fruit on them found in the permafrost.

For several hundred years after the flood, volcanos would continue to erupt. There would be a great deal of earthquake activity more frequently then now both on land and on the ocean floors causing tsunamis more massive than what we have seen recently. The volcanic activity, earthquakes, tsunamis, hurricanes and other severe weather we experience today are all a testament to the fact that the earth is still seeking equilibrium after the massive global flood. The breaking up of the earth's crust during the explosions of the fountains of the great deep and then the collapsing of the ocean floors that shoved up the continents with their mountain ranges left faults in the crust all over the earth that are still shifting today. Rainfall and the moving of air streams in the atmosphere caused some areas to become very fertile while others, over time, turned to deserts. Even massive lakes evaporated away leaving nothing but sand and only the hardiest of plants and animals who had the ability to adapt and survive these most hostile desert environments.

According to the providence of God, Noah and his family landed on a mountain that was not far from the fertile land of what is now known as the Mesopotamian Valley. In that land southeast of Ararat, mankind would find a place where they could flourish for four generations after Noah before being sent out into the rest of the earth as a result of God's confusing their language. As we will see, the very first culture that was established was in the land of Shinar, also understood to be Sumer. It has already been stated that Sumer (Shinar) is known to be the oldest culture in human history. It is by no coincidence that the Bible records the first city built after the flood was Babel in Shinar. That is exactly what you would expect to find if the Bible is true.

TABLE OF NATIONS

Genesis 10 is also known as the Table of Nations. We are told in Genesis 11:10 that Shem was the recorder of the lists given in the Table of Nations. This is to be expected on the basis of Noah's prophecy suggesting that Shem would be the son who would continue the traditions of the one true God and be the keeper of God's progressive revelation during his generations. We find that Shem was able to maintain contact with only the seven sons of Japheth who were Gomer, Magog, Madia, Javan, Tubal, Meshech, and Tiras. Of those seven sons, Shem only recorded the three sons of Gomer, and the four sons of Javan. The best explanation of this short list is Japheth and his sons moved west into Greece, Italy, Europe, Russia, Scandinavia, and England very quickly after the dispersion and Shem lost contact with them early on. Dr. Bill Cooper has done extensive studies of the Table of Nations.

His thesis was that if the table is true, then the names of the people groups in the lists of Japheth, Ham, and Shem's descendants would be traceable from modern times back to these original families and tribes. After twenty five years of investigation, Dr. Cooper was able to verify 99% of this list as being historically correct. In his book, *After the Flood*, Dr. Cooper gives an extensive report on his research concerning the descendants of Japheth. He has chapter after chapter that records the long lists of kings and tribal leaders in early Europe, including Scandinavia, and England that go all the way back to Japheth. One of the important points he makes in his book is the fact that modern scholarship has tried to either cover up the fact that these records are available or cast doubt on their historical authenticity. They do this cover up because these records confirm the Biblical account and bring into question the present day secular interpretation of history based on evolution ideology. So much for honesty in scholarship.

In the last two chapters of *After the Flood*, Dr. Cooper reports on several of the old Anglo-Saxon and Beowulf stories from Denmark that clearly support the Biblical view that dinosaurs were on the ark and survived for a period of time after the flood. Not only does he repeat old stories but shows pictures of drawings and carvings that depict dinosaur like creatures. There is a picture of a carved Saxon shield that replicates a flying reptile at rest called, Pterodactyl. Another Saxon relief pictures a large bipedal legged animal with small arms attacking large long necked quadrupeds. The relief represents a strong resemblance to a tyrannosaurus rex attacking some apatosaurus. In the Beowulf story, the hero is able to kill a juvenile monster called, Grendel.

The monster had a long tail and stood on two large hind legs while its arms were very small in relation to the rest of its body. Beowulf fought the monster by clinging close to the animals belly so it could not bite him with his large mouth full of sharp teeth. He finally managed to tear one of the beast's small arms off causing it to run to its lair and bleed to death. This was another description resembling a tyrannosaurus rex in ancient folklore.[43] This is consistent with the description of dinosaurs in Job. You would expect to find this kind of evidence if God created the dinosaurs, brought them to the ark, and allowed them to survive for a time after the flood.

Shem's list of Ham's descendants (Genesis 10:6-20) was much more extensive than that of Japheth's. The major reason would be that many of Ham's descendants continued in fairly close contact with the descendants of Shem. Ham's sons were Cush, Mizraim, Put, and Canaan. Cush (Ethiopia) has five sons listed, Seba, Havilah, Sabtah, Raamah, and Sabtecha, while Raamah's two sons Sheba and Dedan are added to the list. Cush's sixth son, Nimrod, is listed separately. Apparently, much was said about Nimrod because he became the leader that won fame and power due to his hunting skills. It seems he used that power to convince the other families to build cities under his leadership rather than to spread out and fill the earth as God had commanded. Nimrod founded all the cities, including Babel, that became the land of Shinar (Sumer) and later moved north to establish Nineveh and other cities that became Assyria. A city discovered in Assyria that existed during the time of the Israel kings was actually called Nimrud, suggesting roots going back to Nimrod. There is some question as to whether Asshur (Genesis 10:22), Shem's son, founded Ninevah.

A more reasonable interpretation is Asshur may have come to that area first but Nimrod took it over and added it to his kingdom. With the growth of animal populations and especially the larger dinosaurs and carnivorous beasts, it would be expected that a hunter who could clear out lands of post-flood dangerous animals would be elevated to hero status. This hunting that led to power probably helped lead to the extinction of some of the larger beasts that made good trophies. Some of these beasts we have already seen in Job called behemoth and leviathan.

Of Ham's other sons, Put does not have any sons listed. He may have died without children. Mizrium (Egypt) has seven sons listed, but of those seven only Casluhim was singled out to identify him as the father of the Philistines. These would become important to Israel's history later. They were people who lived in the land of Canaan who were not descendant from Canaan. Some say they migrated there from islands in the Mediterranean Sea. It seems that Ham's descendants moved west into Canaan, Africa, and east into Asia, China, Japan, Mongolia. They continued to migrate over a land bridge of the Bearing Straights into the Americas. Dr. Henry Morris, in his book, *The Genesis Record*, goes into great detail tracing all of these family's lineages in chapter 10.[44] Dr. Morris also includes these detailed lists in his The *Defender's Study Bible.*

Shem then gave the names of his own family. In chapter 10:21, he mentions that he is the older brother of Japheth. In chapter 9:24 we learned that Ham was Noah's youngest son. In this verse it is not clear if the Hebrew reads that Japheth or Shem was the older. Some assume Shem was the eldest, while others accept Japheth as the eldest.

Either interpretation does not influence the validity of the account one way or the other. Shem also makes the point that he was the father of Eber, even though Eber was his grandson, the son of Arpachshad, Shem's son.

This indicates the importance of Eber in the history of Shem's family. Many scholars believe the name, Eber, was the derivative from which the name Hebrew originated. Eber's cousin, Uz, the son of Aram, the son of Shem, was important because Job lived in the land of Uz, suggesting other tribes continued to worship the one true God for a time besides Eber's descendants. It was Eber who had two sons, Peleg and Joktan. Peleg means "division" and we are told he was named this because he was born in the day the earth was divided. Some interpret this to mean the splitting of the continents after the flood.

There is a lot of evidence to make this interpretation plausible. Peleg more readily seems to refer to the division of families by languages that dispersed into the whole earth out of Babel. Peleg was also the descendant from whom Abraham came, who would father the nation of Israel. We see in Shem's list that he was mainly concerned to list those families who continued to worship the one true God, just as we find in the Genesis 5 list.

It is interesting to note that in Shem's first list, he gives the sons of Joktan only. In Genesis 11, he only lists the sons of Peleg and his descendants. He also includes their ages, just as Genesis 5 does. These listings of ages continue once again through the end of Genesis giving a fairly reliable time line from the creation week all the way to Moses in Exodus.

BABEL

The story of Babel begins in Genesis 11. Chapter 10 gave the overview of the major divisions of families that led to the population of the whole earth. Chapter 11 records the events that caused the dispersion of mankind around the world. It gives the background that helps us understand why there are many languages even though we are all descended from the same ancestors. It also gives us the background needed to explain how the many different human physical characteristics developed as a result of the dispersion. More so than these, the dispersion account gives us insight into the events of Pentecost in Acts 2.

Genesis 11:1-9 begins by telling us that after the flood, all humans used the same language (verse 1). This was used to introduce what was to come and make it clear that all languages are the result of an act of God alone. It also infers that all of mankind used the same language before the flood. The same language given to man the moment Adam was created. We can also infer that, in the new heaven and earth, we will all speak the same language once again. The power of speech is one of those God given abilities that separates man from all animals and confirms man's status as being made in God's image. Language was not something that evolved as man grew in intelligence. It is a God given trait that made it possible for God to communicate with man and allow man to be responsible for what he was told. Language is also a wonderful gift that allows us to communicate to God our deepest emotions, our most debilitating fears, our hopes, our faith, our gratitude, our love. Most of all, it is our only avenue of salvation.

Romans 10:17 makes this clear, "So faith comes from hearing and hearing by the word of Christ." It is through language that we arrive at faith in our hearts that saves us. The most important words we can speak are those describing the gospel of Jesus Christ. Those words have saving power.

Genesis 11: 2-4 has given us a window of insight into what took place after Noah and his family left the ark. We are told they journeyed east and found a plain that was to become Shinar. We were introduced to Shinar in the Table of Nations (Genesis 10:10) that became the kingdom of Nimrod. In this kingdom, he built four cities, the first being Babel; the same city mentioned here in chapter 11. We have already seen in the account of Noah cursing Canaan, in Genesis 9, that that event took place long enough after the flood for Noah to grow a vineyard and Ham to have produced his fourth son, Canaan. Cush was Canaan's brother and he, Cush, was the father of Nimrod. Counting Noah, Nimrod would have been the forth generation after the flood.

Counting Noah, Peleg was the sixth generation after the flood. If the division of the nations took place when Peleg was born, as scripture suggests, then according to the chronologies given in chapter 11, the dispersion took place 99 years after the flood in the same year Peleg was born. Verse 4 tells us that they started to build Babel to keep all the families from being "scattered abroad over the face of the whole earth." That would mean that all of the families of up to six generations from Noah on made up the city. Genesis 11:10 and following tell us that each of the fathers in the descendants of Shem had other sons and daughters besides those listed in Shem's record. That would be true of the other brother's families as well. All this suggests that the population of Babel would have been no more than a few thousand people if not only hundreds.

The ages of Shem's descendents given in Genesis 11:10 and following allow us to make some very interesting deductions. First, the ages remain above 200 years and go as high as 500 years (in the case of Shem) after the flood for eight generations. Noah himself lived another 350 years after the flood. When we come to Nahor, Abraham's grandfather, the life spans began to decrease rapidly until we get to Joseph, who lived 110 years. This suggests that there was a change in the environment that, as man's genetic make-up began to weaken under the post flood changes, their life spans decreased as well. Second, all the patriarchs from Noah to Terah, which included men like Shem, Eber, and Peleg, were all still alive when Abraham was born.

As a matter of fact, Shem, Salah, and Eber were still alive when Abraham died. Third, Aram, the sixth son of Shem, was the father of Uz. Job lived in Uz apparently as a descendant of Uz suggesting that these same men who out lived Abraham (Shem, Eber, and Peleg), were also still living at the time of Job's story. These may be the very men that Eliphaz referred to when he said in Job 15:10, "Both gray-haired and the aged are amongst us, older than your (Job's) father." Fourth, all this tells us that during the time of the Genesis account, after the flood, including Isaac and Jacob, there were eye witnesses to the flood.

There were at least witnesses, who knew Noah and Shem, available to all of mankind through the time of Joseph. This statement includes the Pharaoh of Joseph's day. Let me reiterate, this would explain why flood stories continue to this day in almost every culture worldwide.

The forefathers of all these nations knew the flood story very well and the men who experienced it first hand.

Getting back to Babel, Noah, Shem, Ham, Japheth, and all their children were still alive and witnessed the miracle of God confusing the languages. It may be that Eber was given special attention in Shem's account because they chose to not cooperate with Nimrod and the others' scheme to rebel against God. The gist of their scheme was to build a city with a tower in the middle of it that is now known as a ziggurat. Ancient Babel has actually been excavated where one such ziggurat was discovered. It was a pyramid that was built in tiers, one on top of the next, until the very top tier could be used as an observatory of the stars. Many understand this to be the beginning of the worship of the constellations rather than God. Similar structures to these ziggurats are also found in Mexico and South America built by the Aztecs, the Incas and the Myans. Their similar shape and use suggest that they originated in Babel and then were reestablished after the descendents of Indians migrated from Asia into the Americas.

Nimrod's rebellious scheme got God's attention to the point that God Himself came down to see what was in the making. It needs to be restated that Nimrod was the grandson of Ham, and the nephew of Canaan. Chapter 9 told us the story of Ham's disrespect for his father. That attitude seems to have been passed on, not only to Canaan, but to Nimrod as well. Power causes men to do strange things. After Solomon died, his son Rehoboam split the nation of Israel by following the bad advice of his young peers rather than the older and wiser leaders of his father's court. The ten tribes of the north rejected Rehoboam as their king and made Jeroboam their king. Jeroboam became afraid that he would lose his power if the people of the northern tribes continued to go to Jerusalem to worship God.

So Jeroboam made two golden calves (Aaron only made one) and put one in Bethel and one in Dan. He then told the northern tribes that the calves were their new gods so they would not have to go to Jerusalem anymore to worship. This was the beginning of the end of the northern kingdom. In much the same way, Nimrod, wanting to build his own kingdom, led the people to build a place for worship of false gods to keep them in his city rather than obeying God and scattering around the world.

God's answer to this rebellion was to confound the languages in order to make it impossible for the tribes to communicate. Not being able to communicate, they would no longer be able to work out their rebellious plan. We do not know how many different languages were established at that time, but Shem's record seems to indicate that all the heads of families which he listed took part in the dispersion. Those family heads probably represent the new languages that caused these tribes to separate. Of course today, those original languages have developed into other derivatives of the originals. For example, there are several Latin based languages such as English, Spanish, French, etc. that have similar roots. Linguists tell us that all the languages around the world can be traced back to root languages from the past, supporting the Biblical record's account of the origins of different languages.

The Biblical record was passed down from Adam to Noah and from Noah to Moses. It would make reasonable sense to accept the possibility that the language of scripture would be the common thread that connects it all. There is the possibility that Eber was also the one who became known as the father of the Hebrews.

This may have been because it was his family that maintained the Hebrew language when God confounded the languages. All this implies that Hebrew would have been the original Biblical language passed down to Moses by Eber through Shem from Adam. Will we all speak Hebrew in Heaven? Who knows, but what we do know is that we will all have the same language once again. How do I know this? In heaven we will all be a single family once again speaking our Father's language.

Genesis 11:6 gives us the reason why God confused the languages, "... and now nothing which they purpose to do will be impossible for them." What was God suggesting here? Was He concerned that man, with all of his intelligence and know how, would be able to finally make his way to heaven on his own? According to the context and understanding of what had recently gone before (the global flood because of man's sin and rebellion against God), the better interpretation would be God's concern for man's immediate slide back into the level of sin that led to God's destruction of the earth. They were building the city in rebellion to God's command to fill the earth, and they were already going back into worship of the creation (themselves) rather than the Creator.

That was what had unleashed their total depravity before the flood. Dividing them up and sending them out into other environments that would keep them busy just trying to stay alive, would slow down the process of man's return to total rebellion. That slow down would give God the time He needed to complete His redemptive work through His coming Son. As we study the history of all those nations who stayed in that region, God did bring judgment on them later at different times. Sodom, Gomorrah, Canaan, Babylon, Assyria, Egypt, Edom, and even Israel were all judged by God at some point in history.

Rather than bringing one worldwide disaster again, God judged them individually over time. The confounding of languages only served to slow down the process of mankind being totally over taken by their own depravity once again.

RESULTS OF BABEL

The most obvious result of the Babel miracle (God's intervention in human history) was the spreading of mankind around the globe. This isolation of family groups because of different languages was a result of those families moving further and further apart. This separation led to the development of many diverse cultures with very unique traditions and lifestyles.

The study of anthropology is not the study of humans evolving out of simple pre-human animals slowly developing better ways to survive as their brains became bigger. Actually, it is the study of humans moving into various environments around the world and learning how to survive in those challenging and very different environments. What we find is that some cultures developed the technologies that they already had at Babel and grew into empires like Egypt, the Hittites, and early Chinese dynasties. Even the descendants of Ham that moved into the Americas give evidence that they used technology they had carried with them from Mesopotamia. Their style of worship and pyramid building were definitely similar.

Some lost technology and developed into less advanced cultures that chose not to build large cities and elaborate economies.

Those of Japheth's clan that moved into the northwest regions of Europe had to adjust to the extreme changes in the weather that, as the glaciers grew further south into those regions making life harsh, slowed their cultural growth until finally being influenced later by the Greek and Roman cultures that remained further south in less challenging climates. The so-called Neanderthal people were a product of the ice age challenges. It would also be reasonable to assume that living in caves became a viable alternative for many family groups, until it became more advantageous to build huts and live in villages. Those who moved into Africa chose to live more simplistic lives and obviously migrated as far as Australia. Eskimos and other people groups in the northern regions developed new technologies to make it possible for them to live in the most extreme of environments.

Another obvious result of the dispersion of family groups into isolated environments was the development of distinctive physical characteristics. The differences in body characteristics, such as skin color, hair texture and color, eye formation and color, bone structure in the face and body, were used by evolutionists to support their theory that not all humans had the same ancestry. This was the cause of the concept of different races. Evolution was a very racist theory at one time, until science proved what the Bible has always said, that all humans came from a common ancestor, namely Adam and Eve, and Noah and his wife. It was a Bible believing Austrian monk and botanist named Gregor Mendel who discovered the principles of genetic characteristics by pollinating peas. He showed that as reproduction becomes more isolated and the same genetic traits continue to be bred into a people group that those characteristics will become dominate in their gene pool.

In other words, Noah's sons and daughters-in-law had all the genetic possibilities in their gene pool.

As their descendants moved away from that mixed gene pool and began to reproduce within their own isolated communities, those who possessed physical traits that became the most accepted in that society were the ones who were allowed to reproduce most prevalently until those traits became dominate in their gene pool. In some places, dark skin, brown eyes, and kinked hair became the most dominate characteristics. In other places, blonde hair, blue eyes, and light skin became the dominate characteristics. Other societies developed olive skin, slanted dark eyes, with dark hair. It was all about genetic isolation, not race.

In our day, we are seeing a new movement in genetic development. Caucasians, Mongoloids, Negroids, Arabs and others are beginning to intermarry more and more. As they do so, the genetic make up of their offspring is becoming more and more genetically diversified rather than specific. This diversification is leading to more of a mixture of physical traits in those families. If the trend continues, in time, the human race will once again become less diverse in its physical traits. The genetic pool will return to the way it probably was before the flood, when humans would not have been as isolated as they became after the flood and dispersion. With all this in mind, the point can be made that there is no Biblical prohibitions against mixed marriages. Those prohibitions were a limitation created by racism, not the Biblical worldview. The only marriage prohibition the scriptures make is for believers in the one true God not to marry non-believers.

These unequally yoked unions usually lead believers away from God and cause their offspring to grow up as pagans rather than believers. King Solomon is one example. He had seven hundred wives, most of whom were pagans. The scriptures tell us that these wives turned Solomon's heart away from God (I Kings 11:3). All in all, the Biblical record of the dispersion linked with the facts of genetic heredity is a perfect explanation for how the physically diverse human race has become what it is today. There is only one race and that is the human race descended from either Shem, or Ham, or Japheth. We are all part of the same family and every human is either our brother or our sister in Noah. We enter that family by birth. We become a part of the family of God when we are born again through faith in Jesus Christ as our Savior and Lord. The wonderful thing is, whosoever will may come regardless of who you are or where you come from.

THE COMING OF THE HOLY SPIRIT

It was through the study of the dispersion and the confusion of languages that led me to find answers to some important questions about the meaning of tongues in the book of Acts and the coming of the Holy Spirit. Like all other important doctrinal questions, the foundations to the answers of these questions are found in Genesis. First, why did God use cloven tongues of fire to signify the coming of the Holy Spirit? Second, why was the coming of the Holy Spirit described by two opposite concepts, baptism and filling?

Let us take each question one at a time. First, why cloven tongues of fire were used to signify the coming of the Holy Spirit promised by Jesus in John 16?

The concept of tongues takes us back to the dispersion of mankind into all the world by God confusing man's language in Genesis 11.

In the dispersion account, man was in rebellion against God and so God acted in man's behalf to slow his decline back into being totally over taken by his own depravity. By doing so, God accomplished His objective of man filling the earth once again, in spite of man's rebellion to His command. On the day of Pentecost, we see God doing the reverse. God filled those in the upper room with Himself in the person of the Holy Spirit, and to make it clear that the Holy Spirit had come He gave the sign of cloven tongues of fire resting on each believer. They then began to speak in the various languages of the Jews and proselyte Jews from around the world who were in Jerusalem for the feast. Those listening to the disciples speaking confessed that they heard them speaking in their own foreign languages.

The sign of the Holy Spirit's coming can be broken down into three symbols: divided tongues, fire, and languages. First, the cloven tongues of the *King James Version* seem to refer to that which was equally divided amongst those in the upper room by the Holy Spirit. The message the distributed tongues brought was acceptable to God and came from God. Second, John the Baptist had prophesied that Jesus would baptize the world with the Holy Spirit and fire. The fire would represent the Holy Spirit's task in the lost world to bring fiery conviction of sin, of righteousness, and of judgment. It would also signify the purifying power of the gospel message the Holy Spirit had brought with Him. Third, the different languages would signify that the gospel message was for all nations. The gospel message is the same in every language.

And, just as God had divided mankind by confusing their language at Babel; at Pentecost, God was going to bring all of mankind back together in unity by the gospel message that overcomes all language and social barriers. There are three other clear events that replicated the Pentecost experience.

The preaching of the deacon Philip in the city of Samaria in Acts 8:17, Peter preaching to Cornelius' household in Acts 10:46, and Paul re-baptizing the twelve men who only knew of John the Baptist's baptism of repentance in Acts 19:1-6. In the first instance, it was important to establish that the gospel was for the Samaritans as well as Jews. The Samaritans were looked down upon by the Jews as rejected by God but Jesus had preached to them and prophesied the gospel would be preached in Jerusalem, Samaria, Judea and then to the rest of the world in Acts 1:8. Philip's ministry fulfilled that prophecy.

What is unique about this account of the coming of the Holy Spirit is tongues are not directly given as evidence but it seems to imply the Samaritans had a similar experience to the Jew's experience at Pentecost validating their inclusion as recipients of God's salvation. In the second instance, it was important for new Gentile believers in Jesus to have the same experience as the Jewish believers in order to make it clear that the Holy Spirit was given to the Gentiles in the same way He was given to the Jewish believers. That would make it clear that salvation was available to all of mankind. This was exactly what was accomplished as Acts 11:18 records the Jew's response to Peter's testimony about the event, "... Well then, God has granted to the Gentiles also repentance that leads to life."

The third instance was given to make it clear that baptism in the name of Jesus was a greater baptism than John's.

John the Baptist had a powerful ministry and then was martyred. There was the possibility of a religious sect developing that preached about John and did not preach Jesus. It was important to establish early on that Jesus was greater than John. In each of these accounts the major point is that all groups had received the same Holy Spirit and all had an equal filling through salvation in the name Jesus Christ.

There were several other conversions recorded in the book of Acts. None of those conversions record those people experiencing speaking in tongues. The Pentecost event was given to signify the coming of the Holy Spirit and symbolize what His coming meant. Nowhere in Acts do we find where speaking in tongues became a requirement as a sign for salvation or being filled with the Holy Spirit. Paul actually makes it clear in I Corinthians 12-14 that there is a gift of tongues but not all possess it and that other gifts are more important than tongues. It is not even clear that the tongues he writes of in these chapters are the same as those mentioned in Acts 2. There is the possibility that he refers to a prayer language rather than actual human languages. The coming of the Holy Spirit was given to bring Christians together not divide them again, as at Babel. The real issue concerning salvation is not, have I spoken in tongues, but am I being changed into the likeness of Christ. There is a gift of tongues but like all the other gifts not everyone possesses that gift.

BAPTISM AND FILLING

The second question that was answered for me in Genesis concerning the coming of the Holy Spirit is why was the coming of the Holy Spirit described by two opposite concepts, baptism and filling?

In the gospels and in Acts 1, the coming of the promised Holy Spirit was described by a baptism. John the Baptist said of Jesus in Matthew 3:11, "...He will baptize you with the Holy Spirit and fire." He says a similar thing in Mark 1:8 and Luke 3:16.

Jesus also told His disciples after His resurrection in Acts 1:5, "...for John baptized with water, but you shall be baptized with the Holy Spirit not many days from now." When we get to Acts 2:4, we are told the people in the upper room were all filled with the Holy Spirit rather than being baptized in the Holy Spirit. In Acts and the letters of the disciples, the emphasis became the filling more than the baptism.

Why is this a problem? Baptism and filling are opposite concepts. To be baptized means to be immersed into something. When you are baptized into water, your body is submerged in the water. On the other hand, filling means to have something poured into an object, such as water filling up a glass to the top. The water is in the glass not the glass in the water. Many believe the two words are referring to the same experience and it is only a matter of semantics. That argument never fully satisfied me the more I came to understand that God's word was inspired by God and means what it says. For me, there had to be a broader explanation of these important experiences in the life of a Christian.

I believe the answer to the difference can be found in Genesis 1-3. When God created, His creation was very good or a perfect expression of Himself. More than that, it was permeated with His presence. It was His presence that made it good for only God is truly good.

When man was created, God breathed into him His very own Spirit. Man became a living spiritual being living in a physical body. Adam and Eve not only lived in the presence of God's Spirit, they lived with the presence of God living in them. Genesis 3 records man's rebellion against God and two things happened. He saw that he was naked and he began to hide from God.

What happened? God had told Adam that in the day that he ate of the tree of knowledge of good and evil he would die. We know he did not immediately die physically but these two changes that took place reveal that he did die spiritually. God removed His presence from man and man became evil because, as we have already said earlier in chapter 10, evil is the absence of the presence of God's good. Not only did God remove His presence from man, but because the creation was created for man, He removed His presence of goodness from the whole creation. God had to remove His direct presence (His good, the Holy Spirit) because He is a holy God and cannot be in the presence of sin.

That brings us to Jesus. Jesus promised that after He died on the cross, He would ascend to heaven, sprinkle the mercy seat before God with His blood and then ask the Father to allow the Holy Spirit to return to earth and baptize it once again in His presence. Then, when the Holy Spirit came and baptized the earth once again in His presence, those who had already believed in Jesus and were waiting for the promise of the Holy Spirit were not only baptized in the Holy Spirit's presence but they were also filled with the Holy Spirit as their hearts were already open to His coming. And so, since the day of Pentecost the whole earth has once again been baptized in the presence of the Holy Spirit.

That was a one time but ongoing event that continues to this day. And then, any time a person hears the gospel of Jesus and chooses to accept Him into his heart, the Holy Spirit comes in and fills that person with His presence, just as the disciples were filled on the day of Pentecost. On what grounds do I know this is true? By the teaching of Jesus Himself concerning the coming of the Holy Spirit. John 16:7-15 outlines two ministries of the Holy Spirit. The first is described in verses 7-11.

There we are told that the Holy Spirit will convict the "world of sin, and righteousness, and judgment."

Jesus makes it clear that this ministry is to the "lost world" for it is the "world" that The Holy Spirit convicts of these three needs for salvation that Jesus atoned for. Conviction of sin, of righteousness, and of judgment is the ministry of the Holy Spirit that is described in other New Testament passages as the baptism of the Holy Spirit. On the other hand, in verses 12-15, Jesus described the ministry of the Holy Spirit in the hearts of believers. These believers are those who respond to His conviction ministry (baptism) about Jesus and receive Him as their Savior and Lord. In these verses, Jesus does not speak of the lost world. He speaks to His disciples, who already believe in Him. He tells His disciples, ".. He will guide you into all truth..." The "you" refers to believers only. The second ministry of the Holy Spirit then, is the process of sanctification and the renewing of the minds of all true believers. In other words, the Holy Spirit will grow believers into spiritual maturity after they become Christians. The filling of the Holy Spirit in all believers is the follow-up to His ministry of bringing lost souls into salvation, brought on by His baptism of the whole earth in His presence.

We see these two ministries were expressed in Acts 2 when Peter got up and preached for the first time to the crowd on the day of Pentecost. Peter was filled with the Holy Spirit and stood up to preach without formally preparing a message, and yet, he spoke the truth with power. Why? The Holy Spirit was teaching him what to say as he preached, just as Jesus said He would in John 16:12-15. Those in the upper room, including Peter, were baptized as the Holy Spirit returned to engulf the earth once again with His presence.

Not only were they baptized, but the Holy Spirit also filled them at the same time. Why? Because they were already believers in Jesus as the Christ.

Those in the crowd were all baptized in the presence of the Holy Spirit because the whole world was once again immersed in His presence. The evidence of this baptism is found in Acts 2:37 where it says, "Now when they (the lost people in the crowd) heard this (that they had crucified the Son of God), they were pierced to the heart..." Those words "pierced to the heart" gives evidence of the conviction of the Holy Spirit that leads the lost to salvation. That conviction was evidence of the ministry of the baptism of the Holy Spirit at work on the first day He returned to earth.

We see the baptism at work anytime we are in a meeting where the gospel is preached and souls come under conviction and are saved. We see the work of the filling of the Holy Spirit as Christians are being changed by the Holy Spirit's renewing of their minds and causing them to grow into the likeness of Christ. The world is now baptized once again in the presence of the Holy Spirit at all times, just as it was before the fall.

But, Paul tells us in Ephesians 5:18 that, as Christians, we should be continually being filled with the Holy Spirit. Paul wrote this because he knew Christians can grieve or quench the Holy Spirit when we sin. But, we can be refilled as we obey I John 1:9 by confessing our sins and receiving forgiveness each day. Being continually filled with the Holy Spirit goes on in our hearts each day as we grow in our relationship with Him and continually surrender to His leadership.

GIFTS AND FRUITS

The filling of the Holy Spirit brings up an important question. What is the true expression of the filling of the Holy Spirit? Is it the gifts of the Holy Spirit or is it the fruit of the Holy Spirit?

The gifts and the fruit, like baptism and filling, are two different works of the Holy Spirit. Let us look at each one separately, starting with the gifts. Paul gives a thorough explanation of gifts in I Corinthians 12. First, the gifts are given by the Holy Spirit to Christians when they get saved and when they ask for other gifts. Second, these gifts are given to help Christians strengthen one another as we struggle to be faithful in this fallen world. Paul tells us, in I-Corinthians 12:7, that the gifts are for the body of Christ (all Christians) not the person the gift is given to. Third, the gifts are temporary because they will not be necessary in the new heaven and new earth, where there will no longer be the struggle Christians endure in this fallen world. Fourth, a Christian can be living in sin and not be filled with the Holy Spirit but still minister his gifts. We have seen television evangelists preach the gospel and hundreds of people come to Christ for salvation, only to learn later these gifted men were sleeping with prostitutes.

Pastors will resign from highly successful ministries because of adulteress relationships or misuse of funds. How can this be? The gospel is powerful no matter who preaches it. God's word does not return to Him void. The Holy Spirit also honors His gifts, even when His vessel is not honorable, because He cannot deny Himself. What all this means is, that just because a person is expressing the gifts of the Holy Spirit does not necessarily mean that person is filled with the Holy Spirit. Fifth, we also need to remember that Satan can counterfeit the gifts.

For example, demonic worship that involves possession often includes some forms of tongues. Then there is the fruit of the Holy Spirit. This fruit is described in Galatians 5:22-23. The fruit of the Holy Spirit is very different from the gifts. Here is why.

First, the fruit (love, joy, peace, patience, kindness, goodness, faithfulness, gentleness, self-control) is like a cluster of grapes made up of many grapes and yet they are seen together as a whole. They define the very person and character of God. The fruit is who God is. Second, it is the fruit that the Holy Spirit expresses through a Christian when that person is dead to self and is allowing the Holy Spirit's presence to shine through. Third, the fruit is the opposite expression of our depravity. We do not produce the fruit of the Holy Spirit naturally. The fruit is the expression of the return of His goodness in us as it was in Adam and Eve before the fall.

Fourth, Satan cannot counterfeit the fruit. His fallen nature is not capable of reproducing any of the nine qualities in the cluster.

Fifth, in the chapters of I Corinthians 2-3, Paul describes three kinds of people: the natural man (lost), the fleshly man (saved, but not filled), and the spiritual man (filled with the Holy Spirit). In I-Corinthians 3:12, he tells us that all Christians will be judged by fire to determine the quality of their service to God, gold, silver, precious stones, wood, hay, straw. The only service that will survive this fiery trial is the gold, silver, and precious stones that symbolize only the service Christians do as spiritual servants through the filling of the Holy Spirit. Sixth, the filling of the Holy Spirit is very practical. Ephesians 5:18 and following tells us to be continually filled with the Holy Spirit and then goes on to show us what that filling should look like.

When the Holy Spirit is filling us, we will have a positive attitude, be thankful, serve others, wives will submit to their husbands, husbands will love their wives, children will obey their parents, workers will do a good job as unto the Lord, and bosses will treat their workers right.

In light of what we see about the differences between the gifts of the Holy Spirit and the fruit of the Holy Spirit, which is the true expression of the filling of the Holy Spirit? Obviously, the answer to our question is the fruit. Not everyone can have all the gifts, but everyone can express the fruit. The fruit being produced in us is what indicates whether God is in control of our lives or not. The best way to express the gifts is through the fruit being produced by the filling of His Holy Spirit.

Often, we will refer to a worship service as a Spirit filled meeting because it was exciting and created an emotional high for us. In response to that, we need to remember what former evangelism professor of Southwestern Baptist Theological Seminary, L.R. Scarborough, said many years ago, " It's not how high you jump in worship that matters as much as how straight you walk when you hit the ground." The true expression of the filling of the Holy Spirit is the evidence of His fruit being lived out in our daily lives.

THE BIBLICAL CREATION WORLDVIEW SUMMARY

The Bible is a book made up of several books written by different men over a period of 4,000 years. What makes it so amazing is the fact that it is a continuing revelation of God working through historical events that are taking us to a predetermined end.

It is a book that gives evidence that the supreme Compiler of the whole revelation knew the end from the beginning. The major miracle of the book is that it is consistent with itself. It is this consistency that gives us confidence to accept it as the truth. We have seen how the events in the time line began with a very good creation that did not include death, but it did include choice.

Man made a bad choice, God removed His goodness, and everything became evil, but God promised a Redeemer. That Redeemer came in the person of Jesus Christ, the Son of God. As our Savior, He shed His blood for the forgiveness of all our sins and was resurrected after being dead for three days. He sent the Holy Spirit, according to His promise, and His church has been established since that time. God then, has been writing history through His church as the on going revelation of His continued work through the believers in His Son.

But, that is not the end of the ongoing revelation. Jesus, before going back to heaven, promised to come again physically and historically. He also sent a final inspired written revelation, called the "Revelation," to confirm His promised return and give more details. The question is, how do we know He will come again?

Remember, Biblical faith is objective. It is rooted in what can be observed. With this truth in mind, we can give two answers to our question. First, we have the historical record of God's fulfilled promise to send His Son the first time. We know Jesus is the Son of God because of His historical resurrection. Christmas is the celebration of the coming of the promised Savior, an objective historical event. Second, God promised to send the Holy Spirit to fill us once again with His presence.

Ezekiel 36:27 gives the promise, "...And I will put My Spirit within you..." Joel 2:28 gives the same promise, "I will pour out my Spirit on all mankind." Notice Ezekiel refers to the filling ministry of the Holy Spirit, and Joel refers to the baptism ministry of the Holy Spirit. God fulfilled the promise of the coming of the Holy Spirit on the day of Pentecost, an objective historical event. These two historical events establish the basis for our faith in the second coming of Jesus Christ.

The fact that God has already kept these two promises in a physical historical context gives us every reason to believe God will keep His promise to send His Son once again in a physical, historical context. This is the best place to end this book because we have come full circle.

We have covered the creation, the fall, God's promise of redemption, God's past judgments (the flood and dispersion), the fulfillment of the redemption promise through the cross of Christ, and the return of the Holy Spirit.

In Christ, we are spiritually restored to being back where we started with God before the fall. The Biblical creation worldview is, first, that man and creation were both created very good. Second, given choice, man was deceived by Satan to distrust God and fell into sin (rebellion against God). Our sin separated us from holy God. Third, Jesus came in history to take the consequences of our sins on Himself and satisfied the law in order that God might be justified in forgiving us when we repent of our sins. Forth, the Holy Spirit has returned to the earth through the cleansing blood of Jesus Christ. We become Christians by being convicted of sin unto salvation by the Holy Spirit through the gospel message of Jesus Christ, first promised in Genesis 3:15.

Fifth, Christians have been restored to the baptism and filling of the Holy Spirit, just as Adam and Eve were before the fall. Sixth, each individual will be ready for the coming judgment based on the choice each person makes for light or darkness during their life time. We make our choice when we accept or reject Jesus Christ as our Savior and Lord. The whole process brings us full circle back to, "In the beginning, God." Salvation is a new beginning with God.

After the coming judgment, there will be a new beginning for the redeemed in the new heaven and new earth. Sadly, Hell will be a continuation of the first creation's curses experienced in darkness separated from God.

Only God's wrath will be known there. Genesis sets the foundation for understanding the New Testament meaning of the first coming of Jesus Christ and the return of the Holy Spirit to His pre-fall position.

Once we have experienced the baptism ministry of the Holy Spirit and the filling ministry of the Holy Spirit in our individual lives by choice, we are ready to face the rest of what is to come: the second coming of Jesus, the millennial reign of Christ, and the final judgment.

Now that we have laid the foundation of the Biblical worldview in Genesis 1-11, the rest of God's word will begin to make sense to you. As you study the whole council of God, you will continue to see how it all fits together as one continuing, historical, consistent, revelation of God's redemption of fallen man.

The Bible is not a book of allegories and myths devised by man to accommodate the evolution worldview where there is no historical fall.

The Biblical worldview is God's redemption of man in a historical time line and a final establishment of the kingdom He intended from the beginning.

That kingdom, being the family of God, will be made-up of those who were created in His image who then chose in this life to have NO OTHER GODS.

CONCLUSION

God's word is not to be taken lightly. God's word is not subject to interpretation by man's philosophies and his ever changing, so called, scientific theories.

God's word is its own worldview. It is God's worldview given to us to direct us into all truth. It gives us the reality of where we came from and where we are headed as a creation under the control of a sovereign and holy God. God's word must rely on its own self interpretation because it alone is consistent with itself as a continuing revelation building from a specific beginning moving to a predetermined end. As the one true worldview, God's word interprets us. It tells us who we are, where we came from, and where we are headed into the future. Any other interpretation besides the literal creation interpretation gives a false understanding of who we are in light of what God's word tells us about who we are in light of the kind of Creator God He is.

God's word is the foundation of true theology, history, science, and philosophy. The first eleven chapters of Genesis are the foundation on which the rest of God's word is built. If you get those chapters right you will get the rest of God's word right. If you get those chapters wrong, you will get the rest of God's word wrong. Every would be scholar of God's word must be ready to accept the responsibility of James 3:1 before venturing into touching these holy things, "Let not many of you become teachers, my brethren, knowing that as such we shall incur a stricter judgment (greater condemnation)."

God gave a very sobering warning to the Israelites before they entered the land of Canaan. God knew that after the Israelites took control of the land they would become enticed by the worship practices of the pagans. This warning to have no other gods, found in Deuteronomy 12:30-32, applies to Christians today, "... and that you do not inquire of their gods, saying, 'How do these nations serve their gods, that I may also do likewise?'

You shall not behave thus toward the Lord your God, for every abominable act which the Lord hates they have done for their gods; for they even burn their sons and daughters in the fire to their gods. Whatever I command you, you shall be careful to do; you shall not add to nor take away from it."

The "pagan" religion of today called secular humanism that was fostered in the Enlightenment that led to modernism (rational relativism) and has degenerated into postmodernism (irrational relativism) has its roots in an evolution worldview. Like the nation of Israel that did not heed God's warning and allowed pagan practices to become their own which led to the dispersion into Babylon, the United States and other once Christian nations are being taken over by "paganism" which will lead to God's judgment.

These "pagan" philosophies based in evolution have led to legalized abortion, legalized pornography, the acceptance of homosexuality as normal, the weakening of the family and other traditional values that God must finally judge. By adding to and taking away from God's word because of the influence of evolution,

God's people are becoming more and more paganized as the logical conclusions of evolution continue to infiltrate our understanding of truth. The only way to stop our slippery slide into evolution's meaningless black hole is to become totally committed to the authority of scripture as a historical revelation of God's redemptive work from Genesis 1:1 to Revelation 22:21. It is my prayer that this book has helped my readers make that commitment.

After every class I teach, my challenge to my students is, "just believe God's word."

REFERENCES

Ackerman, Paul D., 1986, *It's a Young World After All,* Baker Books, Grand Rapids MI.

Barnes, Thomas G., 1983, *Origin and Destination of the Earth's Magnetic Field,* Institute for Creation Research, San Diego, CA.

Baugh, Carl E., 1987, *Dinosaur,* Promise Publishing Co., Orange, CA.

Baugh, Carl E., 1989, *Panorama of Creation,* Southwest Radio Church, Oklahoma City, OK.

Baugh, Carl E., 1999, *Why Do Men Believe Evolution Against All Odds,* Hearthstone Publishing, Oklahoma City, OK.

Batten, Don, ed., 1990, *The Revised & Expanded Answers Book,* Creation Science Foundation, Green Forest, AR.

Behe, Michel J., 1996, *Darwin's Black Box,* Touchstone Book, New York, NY.

Breese, Dave, 1990, *Seven men Who Rule From The Grave,* Moody Press, Chicago, IL.

Cooper, Bill, 1995, *After The Flood,* New Wine Press, Chichester, Eng.

Cuozzo, Bill, *Buried Alive,* 1998, Master Books, Green Forest, AR.

Daugherty, C.N., 1971, *Valley of The Giants*, Valley of The Giants Publishers, Glen Rose, TX.

Dehart, Roger, 2004, *A Student Guide, Icons of Evolution*, ColdWater Media, Palmer Lake, CO.

Denton, Michael, 1986, *Evolution: A Theory In Crises*, Adler and Adler Publishers, Bethesda, MD.

DeYoung, Don, 2005, *Thousands ... Not Billions*, Master Books, Green Forest, AR.

Dillow, Joseph C., 1981, *The Waters Above*, Moody Press, Chicago, IL.

Farrell, Vance, 2001, *The Evolution Cruncher*, Evolution Facts Inc., Altamont, TN.

Gish, Duane, 1990, *The Amazing Story of Creation From Science and The Bible*, Institute for Creation Research, El Cajon, CA.

Gish, Duane, 1977, *Dinosaurs: Those Terrible Lizards*, Master Books, Green Forest, AR.

Ham, Ken and Taylor, Paul, 1988, *The Genesis Solution*, Baker Book House, Grand Rapids, MI.

Ham, Ken, 1987, The Lie: Evolution, Master Books, Green Forest, AR.

Hovind, Kent, *Longevity Chart Adam to Joseph*, Creation Science Evangelism, Pensacola, FL.

Humphrey, D. Russell, *Starlight and Time*, Master Books, Green Forest, AR.

REFERENCES

Huss, Scott M., 1983, *The Collapse of Evolution*, Baker Books, Grand Rapids, MI.

Ivey, Robert L., ed., 2003, *Fifth International Conference on Creationism*, Creation Science Fellowship Inc., Pittsburgh, PA.

Kennedy, D. James, 1984, *The Real Meaning of the Zodiac*, Corral Ridge Ministries, Fort Lauderdale, FL.

Kennedy, D. James, 1980, *Why I Believe*, W. Publishing Group, Thomas Nelson, USA

Mc Dowell, Josh., 1972, *Evidence That Demands a Verdict*, Here's Life Publishers Inc., San Bernadino, CA.

Mc Dowell, Josh, 1977, *More Than a Carpenter*, Living Books, Tyndale House Publishers, Wheaton, IL

Moring, Gary F., 2000, *The Complete Idiots Guide to Understanding Einstein*, Alpha Books, Indianapolis, IN.

Morris, John D., 1994, *The Young Earth*, Master Books, Green forest, AR.

Morris, Henry M., 1993, *Biblical Creationism*, Baker Books, Grand Rapids, MI.

Morris, Henry M., 1974, *Many Infallible Truths*, Master Books, Green Forest, AR.

Morris, Henry M., 1974, *Scientific Creationism*, Creation Life Publishers, San Diego. CA.

Morris, Henry M., 1997, *That Their Words May Be Used Against Them*, Master Books, Master Books, Green Forest, AR.

Morris, Henry M., 1984, *The Biblical Basis For Modern Science*, Baker Book House, Grand Rapids, MI.

Morris, Henry M., 1995, The *Defender's Study Bible*, World Publishing, Grand Rapids, MI.

Morris, Henry M., 1976, *The Genesis Record*, Baker Book House, Grand Rapids, MI.

Morris, Henry M., 1989, *The Long War Against God*, Baker Book House, Grand Rapids, MI.

Morris, Henry M., 1997, *The Scientific Case For Creation*, CLP Publishers, San Diego, CA.

Morris, Henry M. and Parker, Gary E., 1982, *What Is Creation Science?*, Master Books, Green Forest, AR.

Oard, Michael J., 1990, *An Ice Age Caused By The Genesis Flood,* Institute For Creation Research, El Cajon, CA.

Oard, Michael J., 2004, *Frozen in Time*, Master Books, Green Forest, AR.

Perloff, James, 1999, *Tornado in a Junk Yard*, Refuge books, Arlington, MA.

Peterson, Dennis R., 1986, *Unlocking The Mysteries of Creation*, Creation Resource Foundation, South Lake Tahoe, CA.

Ross, Hugh, 1989, *The Finger Print of God*, Promise Publishing, Orange, CA.

Sanford, J.C., 2005, *Genetic Entropy & the Mystery of the Genome*, Ivan Press, Lima, New York, NY.

Sarfati, Jonathan and Matthews, Mike, 2002, *Refuting Evolution*, Master Books, Green Forest, AR.

Schaffer, Francis, 1972, *Genesis in Space and Time*, Inter Varsity Press, Downer Grove, IL.

Sharp, G. Thomas, 1992, *Science According to Moses*, Creation Truth Publications, Noble, OK.

Sire, James W., 2004, *The Universe Next Door*, Inner Varsity Press, Downer Grove, IL.

Slabach, Robert E., 2003, *Creation, the Fall*, and the Flood, Evangel Press, Nappanee, IN.

Snelling, Andrew, 1990, *The Revised Quote*, Creation Science Foundation, Acacia Ridge, Australia

Strobel, Lee, 2004, *The Case For a Creator*, Zondervan, Grand Rapids, MI.

Sutherland, Luther D., 1988, *Darwin's Enigma*, Master Books, Green Forest, AR.

Taylor, Ian T., 1984, *In The Minds of Men*, TFE Publishing, Minneapolis, MN.

Thompson, J.A., 1962, *The Bible and Archeology*, Eerdman Publishing, Grand Rapids, MI.

Van Bebber, Mark and Taylor, Paul S., 1994, *Creation and Time*, Eden Communications, Mesa, AZ.

Veith, Gene Edward, 1994, *Post Modern Times*, Crossway Books, Wheaton, IL.

Wells, Jonathan, 2000, *Icons of Evolution*, Regnery Publishing, Washington, DC.

Wilson, William, *Wilson's Old Testament Word Studies*, McDonald Publishing, McLean, VA

Whiston, William, trans., 1960, *Josephus*, Kregel Publications, Grand Rapids, MI,

Whitcomb, John C. and Morris, Henry M., 1961, *The Genesis Flood*, Presbyterian and Reformed Publishing, Philippsburg, NJ.

Whitcomb, John C., 1988, *The World That Perished*, Baker Book House, Grand Rapids, MI.

APPENDIX

The following is a list of web-sites of creation organizations with further resources:

Answers in Genesis, Ken Ham
www.AnswersInGenesis.org

Creation Evidence Museum, Carl Baugh
www.creationevidence.org

Creation Science Evangelism, Kent Hovind
www.drdino.com

Creation Truth Foundation, G. Thomas Sharp
wwww.creationtruth.com

Evolution Facts, Vance Farrell
www.evolution-facts.org

ICR, Institute for Creation Research
www.icr.org

Unlocking The Mysteries of Creation, Dennis Peterson
www.awesomeworks.com

END NOTES

CHAPTER 1

1 Sire, James W., 2004, *The Universe Next Door*, Inner Varsity Press, Downers Grove, IL, p. 17.

2 Guralink, David B.; Friend, Joseph H., editors, *Webster New World Dictionary*, The World Publishing Co., Cleveland, OH & New York, NY, p. 1511.

3 Matthews, L. Harrison, 1971, *Introduction to Darwin Theory*, J.M. Dent & Sons, Ltd. , London, U.K., p. xi.

4 Darwin, C., 1872, *Origin of Species*, 6th ed., 1988, New York University Press, New York, NY, p. 154

5 Ibid., p. 143.

CHAPTER 2

6 Mc Dowell, Josh., 1977, *More Than a Carpenter*, Living Bible, Tyndale House Publishers, Wheaton, IL, pp. 37- 55

CHAPTER 3

7 DeYoung, Don, 2005, *Thousands...Not Billions*, Master Books, Green Forest, AR, pp. 158-170.

8 Kennedy, D. James, 1980, *Why I Believe*, W. Publishing Group, Thomas Nelson, U.S., p. 37.

9 Morris, Henry H., 1976, *The Genesis Record*, Baker Book House, Grand Rapids, MI, p. 37

10 Veith, Gene Edwards, 1994, *Post Modern Times*, Crossway Books, Wheaton, IL, p. 22.

CHAPTER 4

11 Dillow, Joseph C., 1987, *The Waters Above*, Moody Press, Chicago, IL, pp. 113-134.

12 Baugh, Carl, 1989, *Panorama of Creation*, Southwest Radio Church, Oklahoma City, OK, pp. 41-51.

13 Dillow, ibid., pp. 195-215.

CHAPTER 5

14 Barnes, Thomas G., 1983, *Origin and Destiny of The Magnetic Field*, Institute for Creation Research, San Diego, CA, 132 pp.

15 Gentry, Robert V., "Cosmology and earth's invisible realm," *Medical Opinion and Review*, October 1967, p. 65ff.

16 De Young, ibid., pp. 82-97.

17 Ibid., pp. 66-78.

18 Sutherland, Luther D., 1988, *Darwin's Enigma*, Master Books, Green Forest, AR, p. 94.

CHAPTER 6

19 Humphreys, D. Russell, 1994, *Starlight and Time*, Master Books, Green forest, AR, p. 137

20 Morris, Henry M., 1974, *Many Infallible Proofs*, Master Books, green forest, AR, pp. 337-342.

21 Kennedy, D. James, 1989, *The Real Meaning of the Zodiac*, Coral Ridge Ministries, Ft. Lauderdale, FL, p. 8

CHAPTER 7

22 Wald, George, "The Origin of Life", *Scientific America*, vol. 191 (2), August 1954, p. 46.

23 Gish, Duane, 1977, *Dinosaurs: Those Terrible Lizards*, Master Books, Green Forest, AR, p. 55.

24 De Young, ibid., pp. 46-62.

25 Schweitzer, Mary H., et. al., *Science*, vol. 307, no. 5717, pp. 1952-1955, March 25, 2005.

26 Boswell, E., *Montana State University News Service*, March 24, 2005.

27 Adler, J., *Time*, November 28, 2005, p. 53.

28 Wells, Jonathan, 2000, *Icons of Evolution*, Regnery Publishing, Washington, DC, p 72.

29 Kennedy, ibid., p. 70.

CHAPTER 8

30 Taylor, Ian T., *In the Minds of Men*, TFE Publishing, Minneapolis, MN, p. 368.

31 Williams, George C., "*Sex and Evolution*," Princeton, NJ, Princeton University Press, 1977, p. 201.

CHAPTER 9

32 Wilson, William, *Wilson's Old Testament Word Studies*, MacDonald Publishing, McLean, VA, p. 6.

CHAPTER 10

33 Morris, Henry H., 1995, *The Defender's Study Bible*, Master Books, Green Forest, AR, p. 11.

CHAPTER 12

34 Sanford, J.C., 2005, *Genetic Entropy & the Mystery of the Genome*, Ivan Press, Lima, New York, NY, p. 151.

35 Whiston, William, trans., 1960, *Josephus*, Kregel Publication, Grand Rapids, MI, p. 27.

CHAPTER 13

36 Morris, Henry H., 1976, *The Genesis Record*, Baker Books, Grand Rapids, MI, p. 181.

37 Ibid., p. 144.

38 DeYoung, ibid., pp. 46-62

39 Morris, Henry, H., 1974, *Scientific Creationism*, Master Books, Green Forest, AR, pp. 137-140.

40 DeYoung, ibid., pp. 110-121.

41 Ibid., pp. 66-78

42 Morris, Henry H., 1977, *The Scientific Case for Creation*, CLP Publishers, San Diego, CA, pp. 55-59.

CHAPTER 15

43 Cooper, Bill, 1995, *After the Flood*, New Wine Press, West Sussex, England, pp. 130-160.

44 Morris, Henry H., 1976, *The Genesis Record*, Baker Books, Grand Rapids, MI, pp. 245-266.